Changing
THE FACE OF
THE WATERS

Changing
THE FACE OF
THE WATERS

The Promise and Challenge of Sustainable Aquaculture

THE WORLD BANK
Washington, DC

ISBN-10: 0-8213-7015-4
ISBN-13: 978-0-8213-7015-5
eISBN-10: 0-8213-7016-2
eISBN-13: 978-0-8213-7016-2
DOI: 10.1596/978-0-8213-7015-5

Cover photo: Michael Phillips and Sena S. DeSilva, NACA.

Library of Congress Cataloging-in-Publication data has been applied for.

CONTENTS

BOXES, FIGURES, AND TABLES

Boxes

Figures

Tables

PREFACE AND ACKNOWLEDGMENTS

The World Bank Group has already recognized the important role of aquaculture, investing approximately $1 billion in aquaculture projects or projects with an aquaculture component. As with any rapidly evolving industry, there are challenges across a spectrum of policy, social, and technical issues. The challenge of sustainable aquaculture is to contribute to national objectives for economic development and food security while simultaneously addressing poverty reduction and environmental protection.

This study provides strategic orientations and recommendations for Bank client countries and suggests approaches for the Bank's role in a rapidly changing industry with high economic potential. It identifies priorities and options for policy adjustments, catalytic investments, and entry points for the Bank and other investors to foster environmentally friendly, wealth-creating, and sustainable aquaculture.

The audience to which this study is addressed includes client countries' policy and decision makers in aquaculture, fisheries, and natural resource management, as well as individuals addressing poverty issues, agriculture development, and environmental protection. The target audience also includes food industry and food trade professionals, the scientific community, development partners, and persons engaged in human capacity development for aquaculture.

Aquaculture can be defined as the farming and husbandry of aquatic organisms, such as fish, crustaceans, mollusks, and seaweed, and the production of freshwater and marine pearls and a variety of other aquatic species, such as crocodiles, frogs, sponges, and sea cucumbers. (The word "fish," unless otherwise stated, is used throughout the report in the generic sense to cover all aquatic animal production, including fish, crustaceans, and mollusks.)

The Food and Agriculture Organization of the United Nations (FAO) defines aquaculture as follows:

> The farming of aquatic organisms in inland and coastal areas, involving intervention in the rearing process to enhance production and the individual or corporate ownership of the stock being cultivated.

The definition distinguishes aquaculture from capture fisheries; in fact, the growth potential of aquaculture lies primarily in its fundamental differences from capture fisheries: with aquaculture, far greater control can be exerted over inputs and production.

In 2001, aquaculture was recognized as a separate economic activity under the International Standard Industrial Classification of All Economic Activities. The collection of statistical data on aquaculture, separate from fisheries data, is a recent endeavor in many countries. In this study, we refer to aquaculture as a sector.

The study reviewed current trends in aquaculture and aquaculture projects and programs supported by the Bank and its client countries and other international financial institutions (IFIs) and donors to assess their roles and impacts. Aquaculture codes, guidelines, legislation, and recommended practices were examined, as were the roles of the public and private sectors and the nature of the institutional frameworks for development and management of aquaculture. Two background studies explored Asian experiences in the use of aquaculture for poverty alleviation and in the transfer of technology and human capacity building. Two additional background studies on aquaculture in Sub-Saharan Africa and in Brazil were complemented by literature surveys and discussions. The study drew on previous work and work in progress by FAO, the WorldFish Center (WFC), the Network of Aquaculture Centres in Asia-Pacific (NACA), and the Norwegian salmon industry.

The review of status and trends drew on draft global and regional reviews prepared by FAO as a result of a mandate by the FAO Committee on Fisheries (COFI) and its Sub-Committee on Aquaculture: to "provide a prospective analysis of future challenges in global aquaculture as a basis for a discussion of the longer term direction of the Sub-Committee's work" (para 73) and "work on environmental risk assessment, including species introductions and undertaking a thematic evaluation of social and economic impact of aquaculture" (para 74) (COFI 2005). The study benefited from a range of studies prepared by WFC and from the collaboration among the World Bank and other institutions, including FAO and NACA.

The identification, planning, design, and implementation of some 67 World Bank projects were reviewed, including 30 projects financed by the Bank's concessional (IDA) or public-sector-lending (IBRD) institutions, and 8 projects with an aquaculture component financed by the Bank's private-sector-lending institution (IFC) (see annex 3). The study also examined project evaluations and experiences of other IFIs, including the Asian Development Bank (ADB) and the Inter-American Development Bank (IADB), bilateral and multilateral development assistance projects (for example, FAO and the United Nations Development Programme [UNDP]), and private and public sector investments. Details of the development agency aquaculture portfolios examined are provided in annex 3. Aquaculture components are frequently embedded in projects and programs with a broader scope, for example, coastal management or rural development. The study extracted those lessons of particular relevance to aquaculture that are often lost in evaluations that focus on overall project impact or larger non-aquaculture components.

The analyses used the Project Completion Reports, Implementation Completion Reports, and similar evaluations of the IFIs. Interviews were held with a number of project leaders, task managers, division chiefs, and sector managers within the World Bank Group and IADB. The review also drew on aquaculture evaluations undertaken by ADB, the Canadian International Development Agency (CIDA), the U.K. Department for International Development (DFID), the Danish International Development Agency (DANIDA), the German Agency for Technical Cooperation (GTZ), the International Fund for Agriculture Development (IFAD), the United States Agency for International Development (USAID), and FAO.

The study draws on parallels in other sectors, such as science and technology, livestock, and agriculture, while forging links to generic policy and planning exercises such as Poverty Reduction Strategy Papers (PRSPs) and Country Assistance Strategies (CASs). In the case of Latin America, the study focuses on Brazil as illustrative of many of the issues facing aquaculture on the continent. Although Chile is Latin America's most important producer because of the unique character of Chilean aquaculture (colder waters and species; for example, salmon, trout, and scallops), this study does not specifically examine aquaculture in Chile.

ACKNOWLEDGMENTS

This study was prepared under the leadership of Kevin Cleaver, director of the Agriculture and Rural Development (ARD) Department of the World Bank, and Sushma Ganguly, sector manager of ARD. The task team included Ziad Shehadeh (consultant), Cornelis de Haan (ARD), Eriko Hoshino (ARD), Ronald Zweig (East Asia Region of the World Bank), and Patrice Talla (Legal Vice Presidency of the World Bank). Kieran Kelleher (ARD) was lead author and task manager.

The team extends its thanks to Mafuzuddin Ahmed (Institute for Fisheries Management and Coastal Community Development), Pedro Bueno (Network of Aquaculture Centres in Asia-Pacific, NACA), Edward Chobanian (consultant), Antonio Diegues (Universidade de São Paulo), and Simon Heck (World-Fish Center) for their valuable input through commissioned studies. Particular thanks is owed to the staff members of the Food and Agriculture Organization (FAO) Fisheries Department, Inland Water Resources and Aquaculture Service (FIRI), for their invaluable collaboration and support, and to Trygve Gjedrem (Akvaforsk Genetics Center AS), Otto Gregussen (Embassy of Norway), Michael Phillips (NACA), Barry Costa-Pierce (University of Rhode Island), Robert Robelus (Africa Region of World Bank), Oliver Ryan (International Finance Corporation), and John Moehl (FAO) for their constructive comments, suggestions, and assistance.

The task team thanks the ARD Management Committee and members of the Fisheries Focal Point for their support and guidance, and Melissa Williams, Marisa Baldwin, Joyce Sabaya, Felicitas Doroteo-Gomez, and Eric Schlesinger of ARD, as well as Daud Khan of FAO, for assistance with logistics and production of the study.

ACRONYMS, ABBREVIATIONS, CURRENCIES, AND UNITS OF MEASURE

AADCP	ASEAN-EC Aquaculture Development Coordinating Program
AAPQIS	Aquatic Animal Pathogen and Quarantine Information System
ACIAR	Australian Centre for International Agricultural Research
ADB	Asian Development Bank
ADCP	Aquaculture Development and Coordination Programme of FAO/UNDP
AIT	Asian Institute of Technology
ARD	Agriculture and Rural Development
ASD	amnesiac shellfish disease
ASEAN	Association of Southeast Asian Nations
ATSE	Australia Academy of Technological Sciences and Engineering
BAP	Best Aquaculture Practices
BCSFA	British Columbia Salmon Farmers Association
BMPs	codes of practice and best management practices
BSE	bovine spongiform encephalopathy
CASs	Country Assistance Strategies
CCRF	Code of Conduct for Responsible Fisheries
CERs	emission reduction credits
CGEP	Code of Good Environmental Practices
CGIAR	Consultative Group on International Agriculture Research
CIDA	Canadian International Development Agency
COD	chemical oxygen demand
COFI	FAO Committee on Fisheries

CSIRO	Commonwealth Scientific and Industrial Research Organisation (Australia)
DANIDA	Danish International Development Agency
DFID	U.K. Department for International Development
DSAP	Development of Sustainable Aquaculture Project
DSP	diarrhetic shellfish poisoning
EIAs	Environmental Impact Assessments
EIFAC	European Inland Fisheries Advisory Commission
ESRP	Environmental and Social Review Procedure
FAO	Food and Agricultural Organization
FDI	foreign direct investment
FIRI	Fisheries Department, Inland Water Resources and Aquaculture Service (FAO)
FSR&E	farming systems research and extension
GMOs	genetically modified organisms
GTZ	German Agency for Technical Cooperation
HAB	harmful algal blooms
HACCP	Hazard Analysis and Critical Control Point
IAA	integrated agriculture-aquaculture
IADB	Inter-American Development Bank
IBRD	International Bank for Reconstruction and Development
ICES	International Council for the Exploration of the Seas
ICRs	Implementation Completion Reports
ICZM	integrated coastal zone management
IDA	International Development Association
IDRC	International Development Research Center
IFAD	International Fund for Agricultural Development
IFC	International Finance Corporation
IFF	integrated fish farming
IFIs	international financial institutions
IFPRI	International Food Policy Research Institute
IIC	Inter-American Investment Corporation
INFOFISH	Intergovernmental Organization for Marketing Information and Technical Advisory Services for Fishery Products in the Asia-Pacific Region
INGA	International Network on Genetics in Aquaculture
IP	Indigenous Peoples Policy
IR	Involuntary Resettlement Policy
IRA	Import Risk Analysis
IRR	internal rate of return
ISA	infectious salmon anemia
ISO 9000	International Organization for Standardization 9000
JICA	Japan International Cooperation Agency
LIFDC	low-income food-deficit country

MDGs	Millennium Development Goals
MRC	Mekong River Commission
MRL	minimum risk level
MTDP	Medium Term Development Plan
NACA	Network of Aquaculture Centers in Asia-Pacific
NACEE	Network of Aquaculture Centres in Central and Eastern Europe
NEPAD	New Plan for African Development
NFEP	Northwest Fisheries Extension Project
NGOs	nongovernmental organizations
NOAA	National Oceanic and Atmospheric Administration
NRC	National Research Council
NTBs	nontariff barriers
OIE	World Organization for Animal Health
PADEK	Partnership for Development in Kampuchea
PCRs	Project Completion Reports
PPAs	*Producteurs Privé d'Alevins* (private fingerling producers)
PRSPs	Poverty Reduction Strategy Papers
RAS	recirculating aquaculture systems
RFID	radio frequency identification
SAARC	South Asian Association for Regional Cooperation
SAPA 2000	Sustainable Aquaculture for Poverty Alleviation
SCALE	SAO Cambodia Integrated Aquaculture on Low Expenditure
SEAFDEC	Southeast Asia Fisheries Development Center
Sida	Swedish International Development Cooperation Authority
SMEs	small and medium enterprises
SPS	sanitary and phytosanitary measures
SSA	Sub-Saharan Africa
STREAM	Support to Regional Aquatic Resources Management
TCDC	Technical Cooperation among Developing Countries
UNDP	United Nations Development Programme
USAID	United States Agency for International Development
WFC	WorldFish Center
WWF	World Wildlife Fund

CURRENCIES AND UNITS OF MEASURE

€	euro	Tk	Bangladeshi taka
ha	hectare	ton/tons	metric ton/tons (1,000 kg)
kg	kilogram		
km^2	square kilometer	US$	U.S. dollars (all dollar amounts are U.S. dollars unless otherwise indicated)
m^2	square meter		
m^3	cubic meter		
P	Philippine peso		
		Y	Chinese yuan

INTRODUCTION AND OVERVIEW

Total world aquaculture production will have reached between 35 million and 40 million tonnes of finfish, crustaceans and molluscs in 2010.

—*The state of the world fisheries and aquaculture 1998*

More than half a decade ahead of these projections, aquaculture production has already reached 45 million tons, providing more than 40 percent of the global food fish supply. As production from capture fisheries stagnates, aquaculture is changing the face of our waters.

FOCUS OF THE STUDY

The objectives of the study are to inform and provide guidance on sustainable aquaculture to decision makers in the international development community and in client countries of international finance institutions. The study focuses on several critical issues and challenges:

- Harnessing the **contribution of aquaculture to economic development,** including poverty alleviation and wealth creation, to employment and to food security and trade, particularly for least developed countries (LDCs)
- Building **environmentally sustainable aquaculture,** including the role of aquaculture in the broader suite of environmental management measures

- Creating the enabling conditions for sustainable aquaculture, including the **governance,** policy, and regulatory frameworks, and identifying the roles of the public and private sectors
- Developing and transferring **human and institutional capacity** in governance, technologies, and business models with special reference to the application of lessons from Asia to Sub-Saharan Africa and Latin America

A SURGING GLOBAL INDUSTRY

Aquaculture lies at a crossroads. One direction points toward the giant strides in productivity, intensification and integration, industry concentration, and diversification in products, species, and culture systems. Another direction points toward the risks of environmental degradation and marginalized smallholders. Yet another direction invites aquaculture to champion the poor and provide vital environmental services to stressed aquatic environments.

The development assistance community has an important role to play in supporting countries as they chart these paths onto a balanced road map for sustainable aquaculture. The vision of sustainable aquaculture demands not only a favorable business climate, but also a governance framework that embraces social objectives and enforces environmental standards. Working with client countries, development assistance can build cross supports and synergies between aquaculture's diverse agendas: the market driven, the environmental, and the pro-poor.

Aquaculture production has continually outstripped projections, and there is little reason to believe that it will not continue to do so. It is inherently more efficient than livestock production. The production chain is shorter and more efficient than for capture fisheries. Moreover, the increasing control over aquaculture production systems is in stark contrast to the faltering management of capture fisheries, for which rising fuel prices are having a disproportionately higher impact on costs. Massive productivity gains are resulting in the falling prices of cultured fish and are extending the consumer base. The scarcity of wild fish creates further market space, while supermarket chains demand stable supplies—uniform-size fish with clear traceability that cannot be readily supplied by volatile capture fisheries. In addition to fish production, responsible aquaculture brings environmental benefits, integrating waste management in urban and rural settings.

New legal instruments are emerging to expand the use of aquatic commons, and good governance is taking firm root in several developing countries. Aquatic farming is likely to intensify, expand, and diversify in ways that currently are unforeseen. Artificial selection may increasingly supplant natural selection and transform aquatic food chains. From fish foods to pharmaceuticals, from ocean carbon sequestration to management of entire aquatic ecosystems—society will progressively extend control over aquatic resources.

This control will drive a gradual convergence between aquaculture and the ecosystem approach to fisheries called for in the Johannesburg Plan of Action.

Fish Supply and Benefits from Aquaculture

Fish supply from capture fisheries has stagnated over the last decade, and no major increases can be expected even with improved management practices and benign climate change. Fish is often the lowest-cost animal protein and the world's growing food fish supply gap has a disproportionate impact on the nutrition and health of the poor. Aquaculture must fill that growing supply gap. It is the world's fastest-growing food production sector, growing at an annual average rate of 10 percent since the mid-1980s, reaching 59.4 million tons (including aquatic plants) with a farmgate value of $70.3 billion in 2004. More than 90 percent of aquaculture production occurs in developing countries, and China alone accounts for 67 percent of global production. Aquaculture products account for about 15 percent of global consumption of fish and meat, and in the near future aquaculture is likely to contribute more than half of the world's supply of food fish. In addition to a growing list of nonfood products and environmental services, aquaculture provides an important livelihood, directly employing more than 12 million people in Asia. The sector provides important foreign exchange to many developing countries, as trade in cultured fish products accounts for 22 percent of the world trade in fish, mostly from the developing world.

Structural Change—From Cottage to Corporation

Modern aquaculture has developed into a dynamic, often capital-intensive business, often with investment by large vertically integrated corporations, some of which are key players in the food retail industry. There have been major increases in aquaculture productivity along the entire production and supply chain. This growth is attributable to a complex interplay of factors:

Technological advances (in particular, improved broodstock and seeds), improved fish nutrition, and better control of diseases, which drive fish prices downward and open new markets

Intensification of most forms of aquaculture, including diversification of culture systems, the species cultured, business models and feed supplies, and integration of aquaculture with farming and waste disposal systems

Consolidation and vertical integration through acquisitions and alliances, contracting of the supply chains, elimination of middlemen, generation of cost savings, and facilitation of improved quality of products in response to national and international market requirements (this structural change alters the distribution of value added along the production and supply chain, and forces small producers to organize or risk becoming marginalized).

The single most important driver of aquaculture is the market, whether for smallholders or larger commercial farms. It drives increased production, further intensification, and competition resulting in reduced product prices and further market penetration—an expanding feedback loop along a productivity timeline.

A Knowledge-Based Industry

The modern fish farm is an intensive knowledge-based enterprise, serviced by dedicated commercial scientific institutions devising new technologies and innovations for corporate clients that move to ever more productive and intensive farming practices. Aquaculture has moved from an art to a science. It has diversified, intensified, and advanced technologically. Improvements in genetics, nutrition, disease management, reproduction control, and environmental management continue to widen choices for aquaculture, improve its efficiency and cost-effectiveness, and optimize resource use. These advances not withstanding, aquaculture is still an infant industry. It lags well behind agriculture in application of science and technology and in value chain productivity. Despite advances in the production of high-value carnivorous species (such as salmon and shrimp), more than 70 percent of farmed fish are herbivores, omnivores, or filter-feeders (such as carp, tilapia, or mussels, respectively).

CHALLENGES AND APPROACHES

Three major challenges confront aquaculture: sustainable economic growth, environmental stewardship, and equitable distribution of benefits. An effective response to these challenges requires a coherent interplay of private investment and stewardship of public goods. By fostering partnerships and providing access to finance and resources, the international community can help developing countries meet these challenges along two intertwined axes of intervention: good governance and knowledge generation and dissemination.

Good Governance and the Creation of an Enabling Environment

An effective governance framework will embrace policies and regulations molded by a clear vision of the future for aquaculture and a road map to realize that vision. The **policy framework** will address issues of equity and strategy, including the following:

- Principles for use and allocation of the public domain (lakes, reservoirs, sea areas, and freshwater supplies)
- A socially required balance between smallholder and large aquaculture
- Coherence with other policies and strategies, such as those on poverty alleviation, industrial development, water and land use, rights of indigenous peoples, or regional priorities

- Environmental sustainability, including mitigation of social and environmental externalities
- Clear definition of the roles of the public and private sectors
- Sector leadership and coordination
- Aquaculture's fiscal regime

Ideally, a **national aquaculture plan and strategy** will mainstream aquaculture into key planning and policy instruments such as Poverty Reduction Strategy Papers (PRSPs), foreign direct investment (FDI) policies, and rural development strategies. It will create space for aquaculture in the physical planning processes and coastal zone and water management plans. A national plan has a vital role in creating an attractive investment climate and interagency coordination, essential to overcome the dynamic nature of an emerging industry where public authority is dispersed across sectors, agencies, and disciplines. A participatory process to prepare such national strategies and plans will build awareness, guide diagnostics, forge a shared public-private vision, and build partnerships among government agencies and with the private sector, producer groups, and nongovernmental organizations (NGOs).

Good governance will draw on **codes of practice and best management practices** (BMPs) to inform and implement policies and plans. Examples of these norms include the Food and Agriculture Organization (FAO) Code of Conduct for Responsible Fisheries (CCRF) and its accompanying Technical Guidelines; the Holmenkollen Guidelines; the World Organization for Animal Health (OIE) International Aquatic Animal Health Code; and other norms prepared with World Bank assistance, such as the International Principles for Responsible Shrimp Farming and the principles for a code of conduct for the sustainable management of mangrove forest ecosystems. Although the application of these codes may raise production costs, the increased returns from healthy and sustainable aquafarms more than justify the costs.

The **regulatory and administrative regime** will draw on the policies to set out the rights and obligations of fish farmers. The regime may specify, among other things, the following:

- Obligation to acquire permits or licenses to establish a farm, based on responsible physical planning for aquaculture, including zoning and safeguarding critical habitats
- Measures to protect the environment, including environmental impact assessments, audits, environmental monitoring (including benchmarking), and internalizing of the cost of environmental impacts
- Control and enforcement mechanisms and penalties or means to redress damage
- Formal processes for stakeholder consultation with adequate provisions for transparency and involvement of NGOs

- Standards for aquaculture practices and animal health and certification systems for the health and safety of aquaculture food products and the quality of seeds and feeds

Private sector investment has dwarfed public investment in aquaculture. A proactive public sector ideally will be a servant of aquaculture and, in addition to being a steward and guardian, will create an **enabling environment** that recognizes the role of the private sector as the engine of growth, innovation, and change. In addition to setting standards and codes, public authorities can establish a progressive fiscal regime, facilitate access to credit (for example, through secure aquafarm tenure), promote trade, and support applied science and capacity building.

Recommendation. The international community can support client countries to improve aquaculture governance through sector diagnostics and stakeholder dialogue leading to national policies and plans for sustainable aquaculture. This road map will embrace and apply the codes of practices referred to above and establish an enabling environment to nurture private enterprise as a vital innovator and engine of sector growth.

Aquaculture, the Environment, and Human Health

Environmental degradation is aquaculture's downside. Aggressive export-driven expansion has frequently caused environmental degradation to lands, waters, and coasts; encroached on the livelihoods of the poor; and alienated commonages. Biodiversity, critical habitats, and human and animal health have been placed at risk through irresponsible aquaculture. However, under increasing regulation and public and consumer scrutiny and by drawing on improved science, many production systems have become more environmentally friendly, reducing their environmental footprint and even contributing to environmental services. Despite consumer confusion from disinformation on the nutritional quality of farmed fish and on the impact of aquaculture on the environment, aquaculture products continue to capture a growing market share.

Aquaculture has domesticated an array of plants and animals in half a dozen phyla. This diversity of species means aquaculture can not only function, but even thrive in degraded aquatic environments and provide a range of environmental services, including waste treatment, water purification, control of human disease vectors, rebuilding depleted fish stocks, and, possibly in the future, carbon sequestration. The expansion of freshwater ponds can result in the spread of human disease vectors, such as snails (bilharzia) and mosquitoes (malaria). Conversely, the stocking of rice paddies, canals, reservoirs, and other public waters can reduce the incidence of human disease.

Today's farmed salmon is a domestic animal with a widening genetic distance from its wild cousins and today, there is more Atlantic salmon in farms than in the wild. Aquaculture's threat to aquatic biodiversity and wild germ-plasm is very real and growing: massive numbers of artificially propagated fish and invertebrates released in the wild reproduce, and currents broadcast their larvae and progeny over vast distances.

Experience from Asia demonstrated the importance of NGOs and trade in promoting environmentally sustainable aquaculture, which reverted to farmers and national authorities in the form of promoting and adopting better practices that reduced or avoided the use of drugs, and advocated water recycling or treatment before discharge, efficient feeding regimes, the use of healthy seeds, and clean ponds. Ultimately, it is the returns to the farmer that influence production decisions. Farmers learned that environmental responsibility made good business sense and that pollution led to the outbreak of diseases; in short, they recognized the link between disease and the environment.

Recommendation. The international community can support an environmentally friendly and healthy aquaculture by providing funding for the public goods dimensions of aquaculture at national and international levels. This includes protection of aquatic biodiversity and wild stocks—a global public good, through reinforcing application of norms and creation of gene banks. It can also support the development of indicators, scorecards, and certification for environmentally friendly and pro-poor aquaculture, and inclusion of NGOs in constructive and transparent partnerships. Further studies are warranted to monitor and evaluate losses of genetic diversity of cultured species.

Pro-Poor Aquaculture

Experiences in Asia provide lessons on pro-poor aquaculture, which, suitably adapted, might catalyze aquaculture in Sub-Saharan Africa (SSA) and in parts of Latin America. In Asia, aquaculture was developed under two models: one in which commercial opportunities have been opened for enterprises; and one in which long-term public support targeted at the poor has generated the necessary critical mass for smallholder aquaculture. The former has largely been driven by private sector initiatives and enterprise; the latter through national policies and programs and external support. The enterprise model has generated growth and employment, often in poor regions. The public support has endeavored to extend that growth to smallholders through policy support, adaptive technologies, knowledge dissemination, and services. These pro-poor approaches varied widely from country to country and included the following:

■ Equitable access to resources. Water and land are the two essential resources. Both may be underused capital assets. Not only is a system of property rights required, but it is an equitable mechanism for their allocation, administration, and enforcement. An example is the rezoning of rice land in China and Vietnam that allowed rice-fish farming and enabled farmers to exit from poverty.

- Use of public waters. Productivity of public waters, such as lakes, canals, and reservoirs, can dramatically increase by stocking fish. In Bangladesh, when coupled with the grant of community or individual stocking/harvesting concessions, this strategy proved to be an effective means of targeting the landless poor.

- Policy bridges to the enterprise model can create the space for the poor to participate in the enterprises. Examples are the concessions and arrangements facilitating contract farming and nucleus estates in Indonesia. In some countries, promotion of small and medium enterprises (SMEs) has indirectly helped the poor through employment creation.

- Integrated design of infrastructure (for example, rural road networks, flood control, irrigation, and drainage) can open market access, help rice paddy and floodplain fish farming, and reduce risks to farmers through diversification of farm production.

- Knowledge and capacity building has proven crucial for large and small aquaculture. It has been delivered through government extension services, universities, service providers (such as feed producers), NGOs, producer organizations, and networks, such as the Network of Aquaculture Centres in Asia-Pacific (NACA).

The poor are also part of the private sector. In Asia, it was recognized that lack of knowledge and capital, higher risks, high opportunity costs for land and water, and access to markets limited small aquaculture entrepreneurs. As a result, new technologies and innovative culture systems have been developed that specifically target Asia's poor. These technologies and approaches have been shared regionally through the networks supported by the international community. Examples include the Genetically Improved Farmed Tilapia (GIFT) initiative financed by the Asian Development Bank (ADB), and the International Principles for Responsible Shrimp Farming prepared with World Bank assistance.

The relative merits of low-trophic-level culture (carps) versus high-end carnivores (intensive shrimp, sea bream) are a matter of ongoing debate. Intensive culture may be highly profitable but has relatively high risks and a high resource budget. In contrast, most low-trophic-level products are typically low-value herbivores or omnivores, yielding modest profits but requiring ample pond or water space—capital that the poor may not possess. Their culture may not be sufficiently viable to lift small producers out of poverty, although it may be highly desirable from a food security and environmental services standpoint. Thus, as demonstrated by the shrimp farms in West Bengal, there is an appreciation that the culture of high-value fish for export may be a more viable strategy than the culture of low-value food fish for local markets. The strategic approaches in Asia also recognized that the benefits of aquaculture accrue along the production chain—there are poor producers and poor consumers—and employment in processing and services may exceed on-farm employment.

Recommendation. The international community can assist client countries to implement a suite of measures and supports for pro-poor aquaculture. These measures may include opening access to public waters for the landless, providing for nucleus estate arrangements as a condition of commercial concessions, integrating public works such as roads and canals with the needs of aquaculture, and providing for extension services and access to finance. It can also promote equitable trade in cultured fish products and address the effects of unwarranted trade barriers on poorer fish farmers.

Knowledge Generation and Dissemination

Aquaculture in Asia has benefited from sustained public support for technology, innovation, and knowledge dissemination under national and regional partnerships. Advances in fish seeds, fish nutrition, and control of fish diseases have been fundamental, and any country aspiring to create a competitive modern aquaculture industry must establish and maintain the required knowledge infrastructure.

Seeds

Advances in seed production have been the springboard of aquaculture. Increasing numbers of species are being domesticated—their reproduction understood and controlled and their reliance on wild seed reduced. Breeding programs such as GIFT, with an internal rate of return (IRR) of 70 percent, have catapulted productivity to new levels. The growth rate of catfish, scallops, and shrimp has increased by 10 percent per generation. The cost-benefit ratio for Norway's salmon breeding program was 1:15. Today, only about 1 percent of aquaculture production is based on genetically improved fish, highlighting the potential for the creation of improved breeds. Dissemination of improved seed is equally important; for example, inbreeding has reduced the growth rate of Chinese carps by 20 to 30 percent. Such improvements, however, must be accompanied by risk assessments and the application of codes of practices and safeguard policies.

Feeds and Fish Nutrition

Availability of feeds is a major constraint to aquaculture in developing countries. Peri-urban aquaculture benefits from the use of local wastes, while a wide range of polycultures and integrated agriculture-aquaculture systems (for example, fish in association with rice, pigs, or ducks) offer feed options for rural areas. The vast proportion of aquaculture production occurs at the lower trophic levels—carnivorous fish account for less than 8 percent of farmed fish production. Aquaculture uses about 56 percent of global fish meal production and 81 percent of global fish oil supplies. Because there are no prospects of significantly increasing fish meal and oil production, and given the rapidly rising prices of these products, there is an intensive search for substitutes and increasing use of yeasts and other sources of essential nutrients.

Disease

Heightened risk of disease accompanies intensification. Following a series of disastrous disease outbreaks, several advances have allowed production to continue to grow. Disease-free strains are being produced—some as proprietary products. Vaccines have been developed and, in the case of Norwegian salmon, have reduced the use of antibiotics to negligible levels.

Dissemination

In Asia, new fish culture technologies and capacity building spread through regional cooperation fostered largely by external assistance, including NACA, Southeast Asia Fisheries Development Center (SEAFDEC), and others. This cooperation was marked by resource pooling, results sharing, and cooperation and trust. Each initiative built on another, which ensured uptake and continuity after project assistance ended. Developing effective producer organizations is another cost-effective approach to increasing knowledge, achieving economies of scale, capturing value and promoting better practices, and gaining access to credit and markets. One knowledge dissemination model, the One-stop Aqua Shop, has been replicated in several Asian countries.

Capacity Building

The experience in Asia shows that the corporate world was again a key to acquiring intangible capital. Companies and producer groups invested not only in training of their staff, but also in research and innovation. Private demand for technical and scientific skills complemented external support for capacity building. With public support, formal, vocational, and informal training built human capacity. Links were established with external centers of expertise and trainers received needed instruction. Development of social capital through civil society dialogue, community-based approaches, and co-management of natural resources all benefited from sustained support.

Recommendation. *The international community can support applied research and innovation on adapting proven technologies to local conditions and build mechanisms to pilot and transfer the innovations and knowledge to farmers. Particular attention should be directed to productivity gains from fish breeding; integrated farming systems; use of land and water unsuitable for other purposes; and commercially viable and environmentally sustainable systems that can be readily embraced by the poor and landless. The international community should continue to support* *institutions, such as WorldFish Center (WFC), to exploit innovations in aquaculture, not only at low-trophic levels and for direct uptake by smallholders, but also in development of local replacements for critical inputs, such as fish oils, in fish health, and the role of aquaculture in waste treatment in both closed and open systems.*

The Challenge of Aquaculture in Sub-Saharan Africa

In contrast to the rest of the world, per capita fish consumption in this region has declined to almost half the global average and, despite suitable natural conditions, aquaculture provides only 2 percent of the region's supply and makes only a minor contribution to economic growth, employment, and foreign exchange. Past aquaculture development efforts have largely failed because of weak institutions, poor access to finance, and a heavy reliance on failing government extension services and seed production. The focus on subsistence aquaculture may have been misguided, because it often lacked the driving force of market demand and impetus provided by commercial reality.

There is evidence of a sea change—urban demand is driving production increases and new commercial producers are capitalizing on export markets opened by capture fisheries. Nigeria, Madagascar, and South Africa are building the critical mass needed for sustainability, while in Malawi integrated aquaculture-agriculture has proven the vitality of new subsistence farming models. African leaders have also recognized this potential, and key elements of a regional approach have been set out in the New Plan for African Development (NEPAD) *Fish for All* Declaration and Action Plan. Sub-Saharan Africa and parts of Latin America can apply the lessons of Asia across a suite of policies and approaches summarized above.

International Partnerships and Finance

Global investment in aquaculture has been estimated at $75 billion in the 1987–97 period, compared with a World Bank Group (the Group includes the International Finance Corporation [IFC]) investment in aquaculture-related projects of approximately $1 billion (more than 90 percent in Asia) in a longer period (1974–2006). Nevertheless, the international financial institutions have made a significant contribution to the development of aquaculture, particularly through capacity and institution building, support for applied research, development of codes of practice, and capital for investment in the production chain.

Recommendation. In consultation with client countries, the international community should develop a set of safeguards and guidelines for investment in sustainable aquaculture that can be applied by the international financial institutions and extended through the Equator Principles. The safeguards and guidelines would be based on existing international codes and best practices. They should be designed to facilitate greater alignment of external assistance, assist developing countries in establishing an enabling investment climate, and secure responsible foreign direct investment (FDI).

Further studies are warranted to acquire a greater understanding of the dynamics of private investment in aquaculture and the means by which public and international support can catalyze private investment in sustainable aquaculture and the transfer of benefits to the poor.

Sustained support should be provided for well-conceived regional knowledge and capacity-building networks that may be established in Africa and Latin America and have a broadly similar role as NACA in Asia.

Client countries and the international community should be made more aware of the entry points for sustainable aquaculture, including natural resource governance, poverty alleviation, integrated coastal and water basin management, and waste management.

Sustained support for human capacity building and for the transfer and adaptation of proven technologies should be an integral part of sustainable aquaculture development programs.

CHAPTER ONE

Trends in Global Aquaculture

The extraordinary growth in aquaculture production is well documented (FAO SOFIA 2005; FAO in press). This section highlights some of the economic, social, and technical dimensions of this still embryonic sector. Supporting statistical information is provided in annex 5.

EMERGENCE OF A GLOBAL INDUSTRY

The recent expansion of modern aquaculture is marked by several key trends and characteristics described below. These trends are expected to continue as the enabling environment for investment in sustainable development expands: as science and technology yield further productivity gains, reduce the negative environmental impacts of aquaculture, and enable man's intervention and management of marine ecosystems to extend.

Productivity

Technical advances are rapidly increasing aquaculture productivity, tracking a path already mapped by agriculture and livestock. More intensive production systems are being adopted and economies of scale are being realized through larger units, at times at the expense of the smallholder producer. Further productivity gains are being achieved through breeding programs and by improving fish nutrition. Communications and structural change are shortening the supply chains as producers interact more directly with retail chains, thus eliminating several layers of intermediaries. Aquaculture is countering resource constraints with knowledge-based advances using fewer resources to produce more at a lower cost. Pressures to internalize negative externalities, such as environmental impacts, may raise costs but lead to a more sustainable industry in the longer term. Corporate giants like Norway's $4 billion Pan Fish and Thailand's Charoen Pokphand Group (turnover $13 billion) represent cutting-edge endeavors of the aquaculture industry.

Diversification

The aquaculture sector may be the most diverse of the food production sectors in terms of species, culture systems, culture environment, type and scale of operation, intensity of practice, and type of management.[1] The number of species being cultured is increasing while enhanced varieties and strains enable further innovations in production systems. From urban fish farms with recirculating water to seeding of the open oceans, aquaculture presents a challenge to the elaboration of national and international environmental standards. This diversity of production systems and selected business models are further described and provided in annex 1 and annex 6.

Key Drivers—Markets and Globalization

Markets are the dominant force driving aquaculture development, adaptation, and innovation. Price competition, changing consumer preferences, new emerging markets, and compliance with environmental and sanitary standards are forcing adaptation and productivity gains in a dynamic global market. New commercial alliances are stripping intermediaries from the production chain as e-commerce, global product standards, and futures markets replace personal contacts. Strong producers and the consumers gain at the expense of less-organized smallholder producers at the far end of extended production chains.

Aquaculture is already a global industry, and developing countries account for 90 percent of global aquaculture production. Fish is the world's most traded food with more than half of world fish trade originating in developing countries, and aquaculture accounts for an increasing proportion of this trade (trade statistics do not distinguish between captured and cultured fish). Price competition between traditional fish producers and increasingly productive fish farmers, mostly occurring in developing countries, is leading to trade barriers and disputes (box 4.4).

Increasing consumer awareness, biosafety, and traceability issues are molding domestic and international markets; and as the yield from capture fisheries falls, the fish supply gap grows, opening further opportunities for aquaculture.

Modern aquaculture is becoming a knowledge-based industry driven by new technologies, intensive production, and highly competitive global markets. As in other industries in a phase of high innovation, developing countries are likely to lag behind without the catalytic actions designed to help them realize the strategic advantages available during a limited window of opportunity.

Environment

Many of the environmental and resource-related concerns in aquaculture reflect a young industry that has grown rapidly in a regulatory void and with a modest underpinning of science. Aquaculture is a minor environmental offender compared with agriculture and other industries. While some concerns are legitimate, others often lack balance or have little basis in science. This has damaged the public perception of the industry and influenced policy. Substantial progress has been made by both private and public sectors to address the negative environmental impacts of aquaculture. A range of codes and BMPs provides clear guidance for environmentally friendly sustainable aquaculture, which in some cases is a net contributor to environmental health (e.g., cultured seaweeds and shellfish can serve as an important nutrient sink).

Lessons from Agriculture

The recent history of aquaculture parallels that of agriculture; however, it is contracted into decades rather than spread over millennia. Just as the forests of Europe were felled for farmland, the aquatic wilds are being converted to aquatic farms at an accelerating pace and scale. Thus, aquaculture simultaneously poses the risks of transformation of entire wild ecosystems and the promise of managed aquatic ecosystems. However, aquaculture differs markedly from agriculture and livestock production in several fundamental ways. Water, the medium of culture, greatly facilitates the inadvertent transmission and spread of wastes, diseases, and genetic material, including introduced species and strains. Unlike terrestrial ecosystems, aquatic, and in particular marine ecosystems, are often more complex and certainly less understood. Aquaculture poses a range of threats to aquatic biodiversity, and control over breeding and reproduction of farmed species is substantially more difficult than in the case of most livestock.

PRODUCTION, MARKETS, AND TRADE

Production

Aquaculture has grown at an annual average rate of 10 percent since the mid-1980s, reaching 59.4 million tons with a farmgate value of $70.3 billion in 2004 (table 1.1). Production of aquatic animals (excluding aquatic plants) for 2004 is reported to be 45.5 million tons (farmgate value of $63.4 billion). Aquaculture accounted for an estimated 43 percent of the global food fish supply in 2004. In comparison, since the 1980s, capture fisheries have averaged an annual growth rate of less than 2 percent, and their contribution to direct human nutrition has actually declined by about 10 percent, partly because of an increase in the proportion of lower-value species, typically used to produce fish meal for animal feed. Approximately one-third of capture fish production is

Table 1.1	Aquaculture Production and Growth in 2004		
	Production (million tons)	Farmgate Value (billion $)	Growth Rate 2003–04 (quantity)
Aquatic animals	45.5	63.4	6.6%
Total aquaculture production	59.4	70.3	7.7%

Source: FAO Fishstat 2005.
Note: Excludes aquatic plants.

directed to nonfood use such as fish meal; in the near future, aquaculture is likely to provide half of the fish used for direct human consumption.

Low-income food-deficit countries (LIFDCs) accounted for 83 percent of production in 2003. Asia leads world production with 89 percent of aquatic animal production by quantity and 80 percent by value in 2003. China[2] accounted for 67 percent and 49 percent of global production by quantity (table 1.2 and A5.3) and value, respectively. In 2004, the top 10 producing countries accounted for 88 percent of production (by quantity).

Since 1970, aquaculture production in developing countries has increased at an average annual rate of 10.4 percent (7.8 percent if China is excluded) compared with a 4 percent growth rate in developed countries (figure 1.1). The uneven growth of aquaculture in developing countries and among continents (see figure A5.5) is due to the lack of tradition in fish farming, technical or institutional difficulties, and weak supportive knowledge and applied research

Table 1.2	Top 10 Producer Countries by Quantity and by Unit Value in 2004						
	By Quantity					By Unit Value	
Country	Million tons	%	US$ millions	%	$000/ ton	Country	$000/ ton
China	30.6	67.3	30,870	48.7	1.01	Australia	6.61
India	2.5	5.4	2,936	4.6	1.19	Colombia	4.61
Vietnam	1.2	2.6	2,444	3.9	2.04	Ecuador	4.59
Thailand	1.2	2.6	1,587	2.5	1.35	Turkey	4.21
Indonesia	1.0	2.3	1,993	3.1	1.91	Chile	4.15
Bangladesh	0.9	2.0	1,363	2.2	1.49	Japan	4.13
Japan	0.8	1.7	3,205	5.1	4.13	Greece	3.77
Chile	0.7	1.5	2,801	4.4	4.15	Brazil	3.58
Norway	0.6	1.4	1,688	2.7	2.65	Mexico	3.27
United States	0.6	1.3	907	1.4	1.50	Italy	3.10

Source: FAO Fishstat, March 2006.

CHANGING THE FACE OF THE WATERS

Figure 1.1   Aquaculture Production by Developed and Developing
            Countries

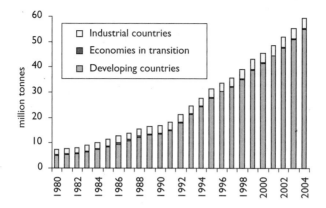

Source: FAO Fishstat 2005.

base. This is the case particularly in Africa and some countries of Latin America, where development potential is high but remains unrealized.

Despite being the focus of much attention, carnivorous fish account for only 7.1 percent of production (table A5.2). Species that feed low in the food chain dominate production (by volume): omnivores/herbivores, such as carps (34.4 percent); filter feeders, such as mussels (35.4 percent); and photosynthetic plants (22.9 percent). Although more than 230 species are cultured commercially, 10 species account for almost 70 percent of all production. Freshwater aquaculture (such as carps) provided the greatest quantity of aquatic animal production (57 percent in 2003) and marine production contributed 36 percent (figure A5.4). Although brackish-water production (such as shrimp) accounted for about 7 percent of the quantity produced in 2003, it contributed 15.3 percent of the value of production.

Employment

Aquaculture production employs more than 12 million people in China, Indonesia, and Bangladesh alone (FAO 2006, in press). Many of these people are rural dwellers and some, such as collectors of wild seed, are among the poorest and most marginalized. Upstream and downstream employment ranges from opportunities at specialized feed mills and genetics laboratories to the aquarium trade and local marketing.

Supply and Consumption

Fish provided more than 2.6 billion people with at least 20 percent of their average per capita intake of animal protein in 2001 (FAO 2004), equivalent to almost 16 percent of the total world animal protein supply. The worldwide average per capita supply from aquaculture has increased from 0.7 kilograms (kg) in 1970 to 6.7 kg in 2003 (an average annual growth rate of 7.2 percent), reflecting an increase in food fish production by more than 500 percent since the early 1980s, compared with an increase of less than 60 percent for meat (excluding milk products) in the same period. Per capita supply in China was 27.7 kg in 2002, compared with the global total of 16.8 kg. Per capita supply in LIFDCs (8.5 kg) is approximately half the global level (table A5.6). In China, more than three-quarters of the food fish supply comes from aquaculture: the share from aquaculture in the rest of the world is considerably lower, 20 percent in 2003, but is increasing.

The share of aquaculture in food fish production has increased from 3.9 percent in 1970 to more than 40 percent in 2004, and approximately 70 percent of total growth in food fish supply since 1985 is attributable to aquaculture (figure 1.2). Since the early 1990s, production from capture fisheries has been relatively stagnant at about 60 million tons of food fish (almost 30 million tons are used for nonfood purposes, including reduction to fish meal). Because of the stagnant supply from fisheries, aquaculture plays an increasing role in determining prices of fish commodities (Lowther 2005).

Figure 1.2   Global Population and Fish Food Supply from Fish Capture and Culture

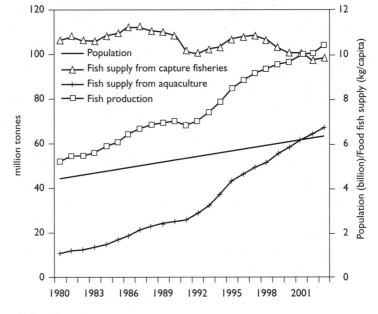

Source: FAO, various years.

Trade

International trade in fish and fishery products has grown from $15 billion (exports) in 1980 to an estimated $71 billion in 2004, and about 37 percent of world fishery production is now traded internationally. Developing countries accounted for 48 percent ($30 billion) of global exports with net earnings of $20 billion in 2004 (see A5.3 for examples). LIFDCs accounted for 20 percent of exports ($13 billion) and imports were $4 billion—export earnings from fish appear to be paying for food imports in some LIFDCs (Ahmed 2004). The developed countries absorbed more than 80 percent of exports.

FUTURE SUPPLY AND DEMAND PROJECTIONS

The rising demand for food fish is driven by population growth, higher incomes, and urbanization in developing countries. With production from wild fish stocks at or near its limits, aquaculture is foreseen as the only major source of additional supplies. In three different scenarios of stagnating capture fishery production, aquaculture output must grow by between 1.4 and 5.3 percent per year to bridge the projected future supply gap (table A5.5) and provide the estimated 70 million tons of food fish required by 2020. Aggregate production targets of selected countries[3] suggest that these global forecasts may have underestimated the future supply of fish from aquaculture and emphasize the key role and potential of such countries as China, Brazil, and Chile.

Asia is projected to continue to produce the bulk of aquaculture output to 2020 and continued expansion is predicted in Latin America, the Caribbean, and Africa, albeit from a much lower base. In many Asian countries, there is a relatively high price and income elasticity for fish consumption (Ahmed and Lorica 2002), suggesting that with the increase in disposable incomes, consumer demand for fish will increase at a higher rate than that of other staple foods such as meat. Rising incomes in China and India, two of the most populous Asian nations and the top two aquaculture producers, are likely to spur the aquaculture industry to meet this rising demand. Already in China, there has been a rapid rise in the production of high-value species—all for the domestic market. Similar changes are likely to occur with increasing frequency elsewhere (De Silva 2001). For example, Vietnam plans to produce 2 million tons by 2010, generating $2.5 billion in exports and 2 million jobs in aquaculture.

Food fish prices in developing countries can decline as more productive culture systems are adopted. In Bangladesh, the Philippines, Thailand, China, and Vietnam, adoption of improved strains reduced tilapia prices by 5–16 percent (Dey 2000). This mirrors similar developments in the price of farmed Atlantic salmon, sea bass, and sea bream brought about by increased production efficiency through genetically improved strains, higher feed efficiency, and more effective disease control. Innovations have a compounding effect—each innovation feeding off another across a broad swathe of technologies, sciences, and production systems, generating further investment in aquaculture technology and innovation. Rising energy prices may place wild fish products at a competitive disadvantage compared with cultured products, and growth in aquaculture, predicated on substantial productivity gains, presents the most attractive scenario for increased supplies of low-value food fish. Furthermore, aquaculture can deliver more homogenous cultured products to markets in a timely manner and, because of its social importance, growing support for smallholder aquaculture may stimulate production.

THE ROLE OF EXTERNAL ASSISTANCE AND THE INTERNATIONAL FINANCIAL INSTITUTIONS

Market-driven private investment has been the engine of growth in aquaculture, and this growth would have occurred irrespective of the external assistance. Global investment in aquaculture has been estimated at $75 billion in the 1987–97 period, while the combined World Bank Group (that is including the IFC) investment in aquaculture-related projects was approximately $1 billion in a longer period (1974–2006).

Between 1978 and 1983, total external assistance to aquaculture development is estimated at $368 million (Josupeit 1985), of which $190 million (52 percent) originated from the three major international development banks (World Bank, ADB, and IADB). During this period, development assistance to aquaculture increased from 8.5 percent to 17.5 percent of the total allocated to the fisheries sector. Between 1988 and 1995, official aid for aquaculture development amounted to $995 million, of which development banks financed 69 percent. By 1995, the development banks dominated, accounting for 92 percent of external funding (FAO 1997b). Asia accounted for 65 percent of the investment commitments (38 percent of the projects); Africa accounted for 16 percent of commitments and about 25 percent of the projects.

The total value of Bank investment in 24 projects with aquaculture components in the period from 1974 to 2006 was $898 million. The value of total loans approved was just over $1 billion[4] (see table A3.1 for a listing of projects) and aquaculture is considered underrepresented in the Bank's portfolio relative to its weight in the portfolios of ADB and IADB (Chobanian 2006). Of greater concern is the skewed nature of the Bank's aquaculture portfolio in terms of both geographic distribution and performance (see table 1.3). Asian countries (greater than 90 percent of loans by value), and China in particular, have received repeated loans for projects that generally have been judged merely satisfactory. Natural disaster (floods) affected one project in China, and the aquaculture credit component of an Indonesian project failed.

In contrast, in Latin America and Africa, issues of internal policy coordination (Mexico) and donor alignment (Malawi) have prevented drawdown of approved loans. Of the three non-Asian projects for which completion reports are available, one (Ghana) was rated satisfactory, another (the Arab Republic

CHANGING THE FACE OF THE WATERS

Table 1.3 World Bank Projects with Aquaculture Components

Region		Asia	Africa	Latin America	Other	Total
Project cost[a]	$ millions	1,052	37	58	2	1,150
	percent	91.5	3.2	5.0	0.2	100.0
Number of projects	Total	23	4	1	1	30
	Ongoing	4	0	0	1	5
	Dropped	0	1	1	0	2
	Completed	21	3	0	0	24
Outcomes	Satisfactory	14	1			
	Partially satisfactory	2	1			
	Unsatisfactory	2	2			
	Not available	3	0		1	

Source: WB projects database, 2006.
Note: For details, see table A3-1.
a. Includes approved loans not drawn down.

of Egypt) was partially satisfactory, and a third (Kenya) was unsatisfactory. In Kenya, the aquaculture component was abandoned when key counterpart personnel were withdrawn. At the appraisal stage of the Mexican project, the government rejected the establishment of an aquaculture fund. While the reasons for project success and failure are complex, it is likely that the distribution of scarce in-house technical expertise in aquaculture shaped both the success and failure of the Bank's portfolio and its geographic composition. The portfolio review uncovered only three instances in which a comprehensive Bank aquaculture sector study[5] led to environmentally sustainable projects. Preparatory studies were at times cursory or poorly conceived and, as a result, some projects were neither well designed nor effectively implemented. In addition to the loan portfolio, through the Consultative Group on International Agriculture Research (CGIAR), the Bank supports the work of the WFC by providing a grant on the order of $1 million per year.

IFC has invested in five projects in the aquaculture sector between 1998 and 2006 and approved $71 million in loans to assist aquaculture development (table 1.4 and table A3.2). Shrimp production accounts for 100 percent of these loans. A review of available outcomes indicates that the projects have had varying success. For instance, IFC financed a pioneer shrimp-farming project in Madagascar in the early 1990s, and by 1998 the farm accounted for 17 percent of the country's total shrimp production (by value) and 11 percent of the employment in the sector. IFC also supported the largest shrimp farm in Honduras following the devastation caused by the 1997 hurricane, providing an additional 1,000 jobs and training in BMPs in shrimp farming. However, the

Table 1.4 Recent IFC Aquaculture Loans by Region

Region	IFC Loans (US$ million)	Percent	Number of Loans
Africa	6.4	9	1
Asia	45.0	63	1
Latin America	20.0	28	3
Total	71.4	100	All shrimp culture projects

Source: IFC 2006.

environmental management performance of a similar IFC-financed shrimp-farming project in Belize was considered less than satisfactory and failed to demonstrate compliance with IFC and Belizean environmental requirements.

Of particular interest is a pipeline project in Indonesia. The shrimp-farming project refinances an existing nucleus estate and develops additional infrastructure. The farm concession covers an area of more than 20,000 hectares (ha), of which 25 percent is used for fishponds and supporting structures (roads and canals). More than 3,100 of the 3,750 ponds are owned and operated by smallholders under a form of contract farming. The production costs under this model are between 15 percent (for *Penaeus vannamei* species) and 99 percent (for *Penaeus monodon*) below the production costs in other IFC-financed projects.

The important roles of several other donors and agencies, including UNDP, FAO, Japan International Cooperation Agency (JICA), and CIDA, are further explored in the discussion on technology transfer and capacity building (see chapter 3, Technology Transfer and Capacity Building, and annex 4, The Regional Framework for Science and Technology Transfer in Asia). The development of GIFT by the International Center for Living Aquatic Resource Management (ICLARM, now WFC) in collaboration with Norwegian scientists showed an estimated economic internal rate of return of 70 percent on ADB's investment.

CHAPTER TWO

Aquaculture, Environment, and Health

THE IMPACT OF AQUACULTURE ON THE ENVIRONMENT

Responsible aquaculture can provide environmental benefits, while unbridled and irresponsible aquaculture can cause a range of adverse environmental impacts (table 2.1).

Physical Alteration of Land and Habitats

Massive introduction of ponds, cages, or rafts intensifies competition for land, leads to loss of esthetic values, and conflicts with other use of aquatic spaces for fishing, recreation, tourism, or navigation. It can also radically alter ecosystem function and highlights the need for integrated coastal zone management (ICZM) and similar instruments for water basin planning.

As a result of the destruction of mangrove forests, shrimp farming has perhaps borne the brunt of the conservationists' criticism of aquaculture (Naylor et al. 2000). But conversion to shrimp farms has accounted for substantially less than 10 percent of the global loss of mangroves (Boyd and Clay 1998). In many areas, shrimp ponds were built on paddy fields, salt pans, or mangrove areas that were cleared for timber, and the conversion of mangrove areas for shrimp culture has all but ceased because of acid soils, high construction costs, and government regulations. In the Sundarbans in West Bengal, there has been a sequential change from mangrove to agriculture (rice paddy) to shrimp farming as a result of rising population density and changing net returns to land use. More recently, land previously occupied by shrimp ponds is now

Table 2.1 Environmental Costs and Benefits of Aquaculture

Negative Environmental Impacts of Irresponsible Aquaculture	Environmental Benefits from Responsible Aquaculture
■ Loss or degradation of habitats such as mangrove systems ■ Salinization of soil and water ■ Coastal and freshwater pollution; for example, contamination of water and fauna through misuse of chemicals and antibiotics ■ Alteration of local food webs and ecology ■ Depletion of wild resources and biodiversity for seed or broodstock ■ Spread of parasites and diseases to wild stocks ■ Depletion of wild genetic resources through interactions between wild populations and cultured populations ■ Impacts of introduction of exotics (deliberate or inadvertent)	■ Agricultural and human waste treatment ■ Water treatment and recycling ■ Nutrient and heavy metal sink ■ Pest control ■ Weed control ■ Disease vector control ■ Desalinization of sodic lands ■ Recovery of depleted wild stocks ■ Preservation of wetlands

Source: Author.

being used for brick production (Chopra and Kumar with Kapuria and Khan 2005). Because of the high salt content of the soil, abandoned shrimp ponds cannot readily be reused for agriculture. Inadvertent salinization of soils surrounding brackish-water aquaculture sites results from the discharge of pond water into drainage canals, and salinization can affect drinking water supplies. For example, low-salinity shrimp farms could be found some 200 kilometers (km) inland from the Gulf of Thailand and an extensive change in land use has occurred throughout the Chao Prayha delta, giving rise to conflicts with rice farmers (Szuster 2003).

Changes in the food web can occur as a result of a high density of fish cages or mollusk culture that alters the composition of the fauna underneath. These altered bottom-dwelling communities may be less effective in cycling sediment nutrients, resulting in deteriorated water quality with negative consequences for fish health and the environment. Norway has well-developed regulations and guidelines to mitigate such effects, including separation and rotation of sites, environmental impact assessments (EIAs), and prohibitions on the use of areas such as enclosed bays or fjords with poor water circulation. A range of other physical effects and interactions can occur, such as reduced water circulation around cages or entanglement of wildlife (U.S. Department of Commerce 2005).

Reduction of Water Pollution

Unlike terrestrial animal husbandry, the chemicals used in aquaculture (fertilizers and drugs) and aquaculture wastes generally enter the water directly. Nutrient-rich water or sludge may be discharged from ponds, while fecal material and unused feeds from cage culture contribute to the nutrient load, which causes plankton blooms, loss of water quality, and mortality of aquatic animals (NRC 1999). Nevertheless, the overall environmental load from aquaculture tends to be comparatively smaller and more dilute and dispersed than waste loads from livestock, industry, or urban centers. Feed additives and drugs applied in aquaculture directly dose the environment, resulting in antimicrobial resistance in nontarget species, and present dangers to human health (see chapter 2, the Impact of Aquaculture on the Environment). Other drugs (parasiticides, spawning hormones) and chemicals (fertilizers algicides, herbicides, oxidants) disturb community structure, toxicity, and impact on biodiversity. There are several approaches to combating water pollution by aquaculture, most of which are detailed in codes of practice and guidelines on best management practices (see annex 2, Selected Codes, Instruments, and Tools for Responsible Aquaculture). Some of these measures are highlighted below.

Regulation

A range of regulatory measures is necessary, including zoning, EIAs, monitoring of farms, and inclusion of environmental stewardship obligations in the conditions of farm licenses. The regulations may mandate reduced stocking densities and reduced water exchange to decrease the volume of effluents and practices for disposal of pond sediments (for example, drying ponds before removal of sediment; use of sediment as soil fertilizer). Rigorously applying fiscal measures (Pongthanapanich 2005), internalizing environmental costs of aquaculture, and applying the "user pays" principle will help mitigate the negative externalities.

Knowledge

The provision of information to farmers on BMPs will improve environmental management of aquaculture and productivity. Improved siting of aquaculture facilities can benefit from sound physical planning, soil analysis, and the use of models, such as waste dispersion and assimilative capacity models. Improved analytical and design tools are required to assess environmental carrying capacity under different production regimes and to design environmentally friendly management regimes.

Integrated Aquaculture

Integrated aquaculture systems can provide efficient and environmentally sound recycling of nutrients and organic materials. For example, polyculture of Chinese and Indian major carps, which account for the bulk of world aqua-culture production, is environmentally benign and can convert agricultural wastes. In open-water aquaculture, filter-feeding species (such as oysters) and economically important seaweeds (such as *Gracilaria*) use suspended matter and dissolved nutrients, respectively. For example, by filtering plankton, mus-sel farms remove nitrogen from the water at a 70 percent higher rate than in surrounding water (Kaspar, Gillespie, Boyer, and MacKenzie 1985). In China, seaweeds and mollusks cultured in proximity to marine cage culture of finfish reduce nutrient loading from the cages or pens. Pond effluents can be treated by constructed wetlands (settling ponds) that can remove three-quarters of total phosphorous and 96 percent of suspended solids,[6] or as in Israel through integration with the culture of seaweeds and mollusks (figure A6.1).

Feeds

Advances in fish nutrition and breeding have contributed to a significant reduction of environmental loads per unit produced (Ackefors and Enell 1994). Modern feeds contain reduced amounts of phosphorus and nitrogen, and the feed conversion efficiency has improved to as high as 1:1 on some Atlantic salmon farms. These feeds are 87–88 percent digestible, reducing the nutrient content of feces.

Certification and Ecolabeling

Consumer demand and pressure from environmental NGOs on retailers are forcing the adoption of environmentally friendly aquaculture. The certification and ecolabeling schemes attest to the character of the production processes and are means of complying with market requirements and adding value to prod-ucts (see chapter 2, Food Safety, Product Quality, and Certification).

Restocking and Introductions of Fish

Lakes, rivers, and seas are regularly stocked with aquatic species to take advan-tage of the natural productivity. The benefits are well known—often accruing to the landless, as in Bangladesh and India. However, these practices may have undesirable environmental side effects, such as the following:

- Disturbance of the existing ecological balance and loss of biodiversity
- Loss of genetic biodiversity of the wild populations
- Transmission or spread of fish diseases

Many of the negative impacts can be avoided or substantially mitigated through the application of codes[7] of practice to preserve ecosystem integrity and conserve biodiversity and genetic diversity (see annex 2, Selected Codes, Instruments, and Tools for Responsible Aquaculture, and chapter 3, Supplying Improved Seeds).

Introductions

The introduction of a nonindigenous species can transform an ecosystem as shown by the introduction of Nile perch in Lake Victoria, which now dominates the lake's aquatic fauna. While the introduction of Nile perch is generally considered to have yielded economic benefits, the arrival of the invasive water hyacinth blocked waterways and access to riparian villages and fishing grounds, causing major economic losses.

Restocking and Ranching

Restocking of public water bodies with fish is an important pro-poor aquaculture strategy that provides a useful community-level entry point. The risks to wild stocks when stocking large numbers of hatchery-produced juveniles to rebuild endangered species and depleted stocks or to enhance the productivity of water bodies can be reduced by stocking juveniles with minimal genetic divergence from their wild counterparts. This can be achieved by using a large number of breeders, and using genetic tags to permit monitoring and adaptive management. Similar issues arise with regard to restocking and ranching. The genetic interactions between the stocked and wild populations are poorly understood, and the extent to which stocked fish simply replace wild fish in landings or add to total fish landings requires further study.

Escapes and Breeding in Cages

An estimated 2 million farmed salmon escape annually into the North Atlantic, and there are reports of successful breeding in the wild and genetic flow from farmed to wild populations (Naylor et al. 2005). The escapees can disturb spawning grounds, engage in unsuccessful reproduction with wild partners, or successfully reproduce and dilute the genetic diversity of the species. These problems are accentuated in highly fecund species, such as cod, which can complete their reproductive cycle while in the ponds or cages, thereby reducing the scope of possible controls. Such events may supplant wild fish and have adverse effects on wild fisheries.

Loss of Genetic Biodiversity

Traits bred into cultured fish often differ from those of wild fish, and interbreeding with escaped farmed fish may result not only in loss of genetic diversity, but also in loss of important survival traits. The risk is greatest for small populations that are already threatened. Concern over escaped species is likely to increase as genetically modified fish are developed for aquaculture. Commercial farming of transgenic fish—genetically modified organisms (GMOs)—is beginning and pilot programs for transgenic shellfish, freshwater fish, and marine fish have already been carried out (FAO 2000).

Disease

Introductions of nonindigenous species can transfer fish diseases; for example, whirling disease, which is caused by a virus that affects rainbow trout. Whirling disease was introduced to North America from imports of brown trout from Europe that appeared healthy, showed no effects of the disease, and in fact were immune to the disease. Another example is white spot disease, a pandemic in shrimp, which has resulted in hundreds of millions of dollars in losses worldwide.

Mitigating Measures

Practical management measures to safeguard the genetic and ecological integrity of wild stocks include the following:

- **Policies.** Preparation of policies on the use of genetically enhanced aquatic species, reflecting the uncertainties associated with their impact on their wild counterparts and applying science-based risk assessment, combined with an adaptive management approach.
- **Introductions.** There are several important measures, such as effective implementation of codes of practice and guidelines, for the responsible use of introduced species (Bartley, Subasinghe, and Coates 1996) and GMOs; risk assessments; and application of the precautionary approach to species introductions (FAO 1995a), including improved quarantine systems and consultation with neighboring countries before introducing nonindigenous species (including GMOs) into transboundary aquatic ecosystems. Many countries already have strict rules on the introduction of nonindigenous species. Unfortunately these rules may not be strictly enforced, and in shared watersheds and in the sea, species can readily move across the national boundaries.
- **Escapes.** A primary emphasis on prevention through improved cage design, anchoring, net management, and guidelines for vessel operation near farms; contingency plans in case of escapes, including fishing to remove escapees as a condition of farm licenses; formal inquiry; improved fish inventory techniques combined with tagging and identification of escapees using genetic markers; establishment of fish farm zones at a distance from wild stocks (for example, mouths of breeding rivers) to reduce potential wild/farmed species interaction; and farming of sterile fish.[8]

Knowledge

The impact on wild populations of restocking and ranching programs, or of the escape of farmed fish, is poorly understood. Informed policies and the application of mitigating measures require capacity building and expansion of the knowledge base about interactions between cultured fish and their wild counterparts. It requires assistance in the adoption of codes, training of geneticists, and enhanced regional cooperation and networking to share scarce skills.

Dependence on Wild Seed

The collection of wild seed for aquaculture affects the wild stock of the target species. Additionally, the fine mesh nets used to collect shrimp seed remove juveniles of a variety of other nontarget species that affect the entire ecosystem. Damage to wild stocks has not yet been quantified, but there is a greater threat to heavily fished stocks and species that have low reproductive capacities. For species under international management, such as bluefin tuna, production from wild seed collection also poses difficulties in monitoring quotas. In addition to the broad collection of wild seed material, "capture-based aquaculture" relies on the capture of mature adults harvested for broodstock. This is usually an interim measure used while techniques are developed for controlled breeding and mass production of juveniles. Currently, commercial sources of seed do not exist for a number of high-value farmed species, including eel, grouper, yellowtail, and tuna, and their culture is totally dependent on the collection of seed from the wild (Ottolenghi et al. 2004; Nash 2005). The total value of farmed production of these species groups is on the order of $1.7 billion.

On the positive side, the practice is labor intensive and can contribute to poverty alleviation and provides alternative livelihood for coastal communities. In Bangladesh, for example, 150,000 people, mostly the very poor, collect shrimp seed for a livelihood. In the absence of other sources of shrimp seed and the creation of alternative livelihoods, these unsustainable practices will continue. Models indicate that the social cost of biodiversity loss from shrimp seed collection could readily be absorbed by farmers in Bangladesh (Kapuria, Nisar, and Khan 2005).

As new species are cultured, capture-based aquaculture is likely to expand, and every effort should be made to reduce dependence on wild seed. Regulations will need to be based on sound policies and an informed assessment of the impact of these practices on target species and the environment. Incentives for commercial hatchery seed production and applied research on seed production for new cultured species is necessary. Alternative livelihoods will need to be created for those engaged in ecologically damaging seed collection.

THE IMPACT OF AQUACULTURE ON HUMAN HEALTH

Aquaculture poses a range of well-known and largely manageable risks to human health. New forms of animal associations and manmade food chains pose risks, but because of the considerable metabolic differences between cultured fish and humans, the possibility of an equivalent of bovine spongiform encephalopathy (BSE) is considered relatively remote. Human health risks arise from pathogens and the spread of disease vectors, chemicals and toxins, and abuse of antibiotics.

Pathogens and the Spread of Disease Vectors

As with all foods, food safety issues are associated with aquaculture products. In addition, aquaculture can contribute to the spread of water- and insect-borne diseases such as schistosomiasis and malaria.

A range of parasitic worms (flatworms, tapeworms, and roundworms), pathogenic bacteria (*Salmonella, Eschericia, Vibrio,* and others), and viruses are common to both wild and cultured species. While it is possible that poorly managed intensive fish culture may render fish more susceptible to disease or be hosts of human pathogens, there is little evidence that aquaculture contributes to increased incidence of these diseases. Many bacteria are already naturally present in the water, or more commonly introduced by contamination from livestock, by human waste, or through postharvest contamination in product handling and processing. Risks are higher in freshwaters than in marine waters and higher in tropical than in temperate climates. For example, cholera pathogens are a natural component of tropical waters and can contaminate mollusks even without their exposure to human fecal material. The level of pathogenic microbes in marine waters has been a constraint in the shellfish industry throughout much of the European Union. Infection of humans with fish pathogens has been documented in the United States and Israel (Bisharat and Raz 1996; Weinstein et al. 1997). Viruses causing disease in fish are not pathogenic in humans.

The main cause of human infection is the consumption of raw or inadequately cooked fish. Most parasitic worms and bacteria are destroyed thorough cooking and in some cases, by storing fish at low temperatures. Epidemiological evidence suggests that the risk to human health is low and needs to be placed in perspective with improved availability of low-cost fish and increasing sanitary and food safety requirements in commercial aquaculture.

Spread of Disease Vectors

Aquaculture provides a range of environmental services related to human health; in particular, the control of hosts and vectors of pathogens, such as mosquitoes and snails. However, abandoned or poorly managed fishponds have been associated with the spread of schistosomiasis and malaria. Ducks raised in association with fish farms and the use of chicken dung as fertilizer in fish farms may contribute to the spread of diseases, such as highly pathogenic avian influenza, but in this case the ducks are the vector—not the fish (Gu, Hu, and Yang 1996; Skladany 1996; Feare 2006).

Chemicals and Toxins

The risks associated with chemicals and toxins arise from several sources: agricultural chemicals, pesticides, veterinary drug residues, and accumulation of other pollutants. In addition, under certain conditions, some cultured species can produce toxins. Poor farming practices have led to major commercial losses when importing countries have placed import restrictions on shrimp and other products from Asia and Latin America as a result of unacceptable levels of antibiotic residues in products.

Pesticides and piscicides (chemicals to kill fish) present a risk to human health, and their use has to be carefully monitored. High levels of polychlorinated triphenyls (PCBs), dioxins, and other contaminants have been reported in farmed salmon (Hites et al. 2004) and attributed to the bioaccumulation in the fish meal feed. Metals from mining, use of antifouling agents, or natural sources can also accumulate in the edible tissues of cultured fish—in particular, of invertebrates. The levels of mercury and organochlorides in fish are a public health concern. In general, farmed fish, such as salmon, are likely to have lower levels of mercury than their wild counterparts because the feed rations are monitored and they are harvested younger (mercury accumulates during the life of the fish). Although there is still some debate, the benefits of eating fish, particularly for vulnerable groups such as young children and pregnant and lactating women, have been shown to far outweigh the potential adverse effects (Eurofish 2004). An increasing number of studies also show a close link between brain function and consumption of fish oils (omega-3 fatty acids) and a growing body of evidence indicates that reduced fish consumption in the population may result in a lowering of the population's intelligence and in antisocial behavior (see box 2.1).

Toxins

Toxins in bivalves—oysters and mussels—are associated with algal blooms, which are occurring with increasing frequency because of eutrophication of coastal waters. These harmful algal blooms (HAB) have threatened the economic viability of bivalve culture in some parts of Europe because of diarrhetic shellfish poisoning (DSP), paralytic shellfish poisoning (PSP), and amnesiac shellfish disease (ASD). The cost of depuration measures such as those required under EC Directive 91/492, is significant to producers, which is an example of a negative environmental impact on aquaculture caused by other resource users. In China, a World Bank project (see box 2.2) is helping to address these issues.

Antibiotics and Drug Resistance

There are several risks related to the use of veterinary drugs in aquaculture: risks to consumers from ingestion of the drugs, or residues, leading to devel-

Fish are a primary source of a group of important nutrients known as omega-3 fatty acids. The average diet doesn't include enough seafood, and recent studies suggest these fatty acids are even more important than had previously been recognized. In particular, the amount of omega-3 in a pregnant woman's diet helps to determine her child's intelligence, fine-motor skills, and propensity to antisocial behavior.

Children of women who had consumed the smallest amounts of omega-3 fatty acids during their pregnancies had verbal IQs six points lower than average, which could have a serious effect on a country's brainpower if this occurrence were widespread. The finding is particularly pertinent because existing dietary advice to pregnant women in some countries is that they should limit their consumption of seafood to avoid exposing their fetuses to trace amounts of brain-damaging methyl mercury. Ironically, that means they avoid one of the richest sources of omega-3s. Other studies indicate that changes in diet over the past 50 years—particularly changes in omega-3 and omega-6 consumption—are an important factor behind the rise in mental ill health in Britain.

Source: The Economist 2006.

opment of antibiotic resistance in human pathogens (Weber et al. 1994); and risks to workers from exposure to drugs. Many countries have strict controls on the use of veterinary medicines and other chemicals in aquaculture. Partly in response to the strict controls by major importers, countries such as Vietnam, Malaysia, Thailand, Sri Lanka, and China have adopted lists of chemicals approved for use, and guidelines for the proper use of chemicals are becoming important components of BMPs. Development of vaccines and disease-free strains and maintenance of sensible stock densities have reduced the use of antibiotics. In Norway, the decreased use of antibiotics (figure 2.1) is a result of good site selection, site rotation (one generation per site); improved production management; effective vaccines; strict veterinary control of all farms; strict rules for movement of live fish; and use of approved medicines (Gregussen 2005).

Food Safety, Product Quality, and Certification

Fish product standards are increasing in terms food safety, traceability, and quality. The introduction of mandatory Hazard Analysis and Critical Control Point (HACCP) requirements for exports to the European Union (in 1997) and to the United States has had substantial impact on trade in aquaculture

Box 2.2   Win-Win Situations for Aquaculture and
the Environment

China. The total cost of the Sustainable Coastal Resources Project is $200 million (aquaculture component $146 million). The World Bank Loan of $100 million is on-lent to participating provinces/prefectures.

The project's objectives are to (1) support the government's commitment to sustainable development of China's coastal resources, (2) reduce pressure on coastal fishery resources, and (3) help improve aquatic product quality. Project components include the following: (1) the design and implementation of ICZM plans in selected areas of four coastal provinces; (2) production and processing of aquatic products that promote and reinforce sound ICZM policies; and (3) several activities to improve aquaculture production (advanced hatchery technology, new environmentally friendly shrimp culture methods that combat the current disease problems, and HACCP training).

The oyster or kelp culture is sited close to the fish cages to provide a natural source of filtration of waste products. In the case of the polyculture ponds, the waste products that collect in the beds of the ponds are removed yearly and used as farm compost.

India. In the wetlands outside the city, a workforce of about 8,000 produce and deliver about 13,000 tons of fish per year to the 12 million inhabitants of Calcutta. This mosaic of traditional ponds, known as *bheris,* was threatened both by the natural process of delta formation and urban sprawl. Preserving some 3,500 hectares of these wetlands as a home to many migrating birds is the lesser of the environmental services rendered by aquaculture. Production from the 300 ponds relies on the 600 million liters of raw sewage that spew from Calcutta every day. This is the city's sewage treatment plant, deploying a natural cascade of water hyacinth ponds, algal blooms, and fish to dispose of the city's human waste. Moves to replace this natural sewage works with a modern plant were resisted. According to Dhrubajyoti Ghosh, "it . . . took three years to explain to official decision makers that a conventional technology can give way to the traditional one." Although guidelines have been developed for sewage-fed aquaculture, clearly social sensitivities may preclude use of human excreta in aquaculture in some societies (Furedy 1990).

United States. In 2005, about 5 percent of the Chesapeake Bay (see http://www.chesapeakebay.net), the largest estuary in the United States, was considered hypoxic, or a "dead zone." As a result of fishing, disease, and habitat change, the oyster population was less than 1 percent of its original size. Efforts are currently being made to increase the oyster population tenfold through a range of strategies, including aquaculture to improve the water quality of the bay.

Source: World Bank internal project reports.

Figure 2.1   Antibiotics and Salmon Production in Norway

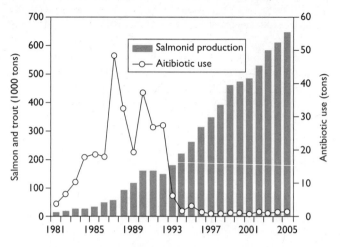

Source: Institute of Marine Research, Norway.

products. Developed countries and a large number of developing countries have instituted HACCP systems, and an estimated 65 percent of the fish products traded internationally are subject to HACCP regulations. Several countries have developed HACCP plans for selected aquaculture products; for example, the United States now has plans for catfish, crayfish, and mollusks. The major exception is Japan, which accounts for about 24 percent of the global fish trade, but which applies different, non-HACCP regulations. The HACCP system is well defined for processing, but with few exceptions, the application of HACCP to the entire aquaculture production chain is not so clearly established. Risk assessment and traceability—the ability to track aquaculture products "from farm to plate"—are becoming integral components of aquaculture management and a fundamental requirement for domestic and export marketing. The traceability content can include a wide range of information (Dallimore 2004), for example, on feeding (to ensure that no GMO feeds are used), slaughter, environmental impact, or even ethical issues, such as worker conditions. The information is transmitted in a wide range of formats, including paper-based reports, bar codes, or radio frequency identification (RFID). Of particular importance has been the traceability of shellfish origin in relation to toxins from algal blooms. Traceability is a key element in certification and ecolabeling schemes, improving the sector's public image and gaining consumer confidence.

Certification

Major retailers seek to differentiate their products through certification (table 2.2). This creates competitive advantage and their requirements may, in fact, be higher than mandated in national regulations or for compliance with the Codex Alimentarius Commission sanitary and phytosanitary measures (SPS).

In response, some producers have undertaken voluntary certification (for example, using International Organization for Standardization [ISO] 9000) for control and marketing purposes. Such certification is increasingly required for access shelf space in the supermarket chains or multiples (see table 2.2). Recent legislation in both Europe and the United States requires mandatory certification to identify whether products are produced from aquaculture or caught wild. Conservation groups, such as the World Wildlife Fund (WWF), and producer associations are working to certify aquaculture products as environmentally friendly and to raise national or regional brand consciousness. Belize, Colombia, and Madagascar have considered using certification so that production from the entire country or state can be differentiated in the global marketplace. Authorities have expressed interest in using certification systems as a basis for issuing aquaculture permits, and investors are interested in a third-party certification as the basis for screening investments. Emerging standards for organic aquaculture will also require such certification processes.[9]

An Increasing Burden

Despite limited capacities and resources, developing countries are making efforts to comply with the more stringent standards. HACCP is increasingly focusing on risk assessment by the operator, and this issue will put further institutional demands on exporting countries (Josupeit, Lem, and Lupin 2001). Some countries have established "zero tolerance" standards for chloramphenicol and nitrofurans based on the minimal detectable limit, essentially precluding trade until the exporting country has made the necessary advances in analytical technology. Each new standard increases the costs for developing countries to comply with such standards and simultaneously creates opportunities for circumventing controls and encouraging corrupt practices, such as the issue of false documentation or failure to inspect shipments. Each new procedure requires human resource development—training of staff both in fish plants and farms and in the services that support the industry. Continued expansion of aquaculture will require collaboration among developing countries to improve their ability to comply with these international obligations and trade standards. The level of hygiene in domestic markets often falls short of appropriate hygiene standards and equally deserves attention.

ENVIRONMENTAL SERVICES FROM AQUACULTURE

Waste Treatment and Recycling

Aquaculture already provides a wide range of environmental services, including waste treatment, recycling of water, and disease control. In the future, this role may become as significant as its role as a supplier of food and other

Table 2.2 Table-Certified Cultured Fish Products in EU Supermarkets

Supermarket Chains	Cora	Inter-marché	Al campo (Auchan)	Carrefour (Italy)	Coop Italia	Métro	Esselunga
Species							
Seabass	✓	✓	✗	✓	✓	✓	✓
Seabream	✗	✓	✗	✓	✓	✓	✓
Certified values							
Full traceability	✓	✓	✓	✓	✓	✓	✓
Flavorsome, tasty	✓	✓	✓	✓	✓	✓	✓
Environmentally friendly	✓	✓	✓	✓	✓	✗	✗
Ecologically/socially acceptable	✗	✗	✓	✓	✓	✗	✗
Other values	Value for money	Convenience				Packaging and freshness	Minimum use of artificial ingredients

Source: Monfort 2006.

Note: ✓ = available; ✗ = not available.

products. The two roles are complementary as nutrients and wastes (for example, from sewage, agriculture, and industry) can be converted to products of value such as fish, clams, or seaweed.

Aquaculture can be a cost-effective natural buffering system for water management. It offers a cleaning and recycling service that is viable in its own right—the clean water is a bonus that opens doors to more sustainable agriculture. Not only are nutrients removed, but also pathogens that may threaten agricultural food production. Pond systems can serve as agents of flood control and water reserves for livestock and gardens. Fish culture can keep irrigation canals free of weeds, rehabilitate sodic lands, and be an integrated part of farm and water management. Rice-fish culture is a proven approach to integrated pest management.

Bivalve mollusks (such as clams, oysters, and mussels) and other filter-feeding animals help to keep bays, estuaries, and lakes clean by "straining" or filtering algae from the polluted waters, while aquatic plants can absorb the dissolved chemicals. High levels of pollution, however, may kill these helpful bivalves, thus compounding the impact of the pollution. Fishing activities add to the pressures on these ecosystems by harvesting the shellfish.

Disease Control and Other Environmental Services

The capacity for disease control is a positive externality of fish culture. Fish feed on mosquito larvae and snails, many of which are intermediate hosts or vectors of human parasitic diseases. Among the most important of the diseases are malaria, filariasis, yellow fever, equine encephalitis, and dengue fever (via mosquitoes); infections by parasitic worms (via snails); and river blindness and sleeping sickness (via flies).

Biological agents, particularly the grass carp, can be profitably and successfully reared in irrigation canals and other freshwater bodies to remove unwanted aquatic weed growth. In addition to the benefits of irrigation schemes, the removal of the vegetation reduces the incidence of diseases borne by mosquitoes and snails (Redding and Midlen 1990). Catfish and tilapia are among the species used to control snail vectors. Rice-fish culture offers a form of benign pest management that reduces the use of pesticides.

The frontiers of aquaculture stretch to marine plankton carbon trading (see box 2.3). Pilot programs have been established that integrate shrimp farming with carbon sequestration using mangroves and other (nonalgae) aquatic plants, such as samphire. The prospects for seaweed farming to contribute to carbon sequestration have not been explored.

Toward Environmentally Friendly Aquaculture

As aquaculture grows, it extends its demands on environmental space, and the governance framework needs to adapt to ensure that the sector makes the transition to responsible and environmentally friendly practices. Sound policies,

About 48 percent of global carbon produced by burning fossil fuels is sequestered in the ocean. Availability of iron limits plankton distribution and growth in many parts of the ocean. Experiments have demonstrated that spreading iron filings in the ocean can result in plankton blooms and large ocean iron fertilization could potentially alter large marine ecosystems, and perhaps even the planet.

Following the Kyoto Protocol, carbon sequestration markets were established to trade certified emission reduction credits (CERs) and other types of carbon credit instruments internationally. In 2006, CERs sell for approximately €25/ton of carbon dioxide equivalents (CO_2e), which suggests that a full-scale plankton restoration program could generate up to €75 billion in carbon offset value. This has attracted investors proposing to fertilize the oceans in exchange for carbon payments. However, such proposals are highly controversial on ecological, technical, legal, and other grounds and reinforce the need for more effective governance of the high seas and greater understanding of the role of the oceans in climate change.

Source: World Bank Carbon Finance Unit (available at: http://carbonfinance.org/); The Global Hub for Carbon Commerce (available at: www.co2e.com); Planktos (available at: http://planktos.com/).

codes of practice, regulatory regimes, and BMPs, including EIAs, physical planning, and economic instruments, are among the tools that can be used (see annex 2). The incentives for environmentally friendly aquaculture are largely commercial: it makes good commercial sense and complies with international trade standards—good environmental practices improve fish health and economic returns. The safeguard policies of the IFIs provide a complementary approach. Consumer awareness, product certification, and technological advances can all contribute—the bottom line is that aquaculture itself needs a clean environment.

As food safety requirements are harmonized at international levels, quantitative risk assessment and traceability are becoming integral components of aquaculture management. Guidelines and measures have been developed to reduce public health risks from pathogens in livestock-fish systems (Little and Edwards 2003) and in sewage-fed fish culture (Buras 1990); potential food safety hazards of products from aquaculture have been reviewed (WHO 1999). The OIE provides guidance on aquatic animal health standards.[10] Improved dialogue and coordination among engineers, public health officials, veterinarians, and aquaculture regulators will improve the environmental services from aquaculture, help reduce health risks, and avoid trade restrictions.

Public investment is required to create a sound governance framework, fill knowledge gaps, and develop human capacity. Although many modern aquaculture systems appear more environmentally friendly and benign than livestock production systems, closer examination of the environmental footprint and impacts of these systems is required. A substantial element in this footprint is a continued reliance on fish oil derived from capture fisheries (see chapter 3, Feeds, Seeds, and Disease) and increased use of energy as production intensifies.

Public Perception and Science

Many of the environmental and resource-related concerns in aquaculture reflect a young industry that has grown rapidly in a regulatory vacuum and with a modest underpinning of science. Scientifically rigorous assessments are required to develop solutions to the many legitimate concerns and to avoid a misinformed public perception of an industry that, at times, has been poorly served by unbalanced science.

CHAPTER THREE

Innovation and Technologies

The number of farmed Atlantic salmon greatly exceeds the number of wild salmon.

—Various sources

Technological advances in fish breeding, fish nutrition, and disease control have resulted in dramatic increases in production and productivity. For example, white leg shrimp output increased from fewer than 150,000 tons to almost 1.4 million tons over a five-year period (table 3.1). Increasing productivity and output has caused a dramatic fall in prices of many major aquaculture products, which, in turn, has attracted new consumers, opened new markets, and made products previously considered luxuries a common item on supermarket shelves. This price decline has not been restricted to the luxury products—carp and tilapia prices have also declined. While a number of techniques and fish breeds are proprietary, many aquaculture technologies are public goods that have given high returns on public and private investment.

FEEDS, SEEDS, AND DISEASE

Fish Nutrition and Feed Supplies

Three categories of fish feeds can be distinguished: (1) natural feeds, such as the plants and detritus grazed by carps in traditional extensive systems and

Table 3.1   Changes in Prices and Production for Genetically Improved Species

Product	Period	Price Decline (%)	Production Increase (%)	
Atlantic salmon	1986/87–2004	20–40	3,108	
Pacific white leg shrimp	Recent	62	854	(2000–04)
Japanese eel	1988–2004	71	159	
Common carp	1984–2004	40	397	
Tilapia	1992–2004	20	164	

Source: FAO Fishstat Plus 2006.

plankton filtered by mussels; (2) artificial feeds, used for species cultured in intensive systems; and (3) waste products, including trash fish and animal and plant wastes. The lines between the feed regimes are increasingly blurred. For example, as extensive farming of tilapia and carp polyculture is being replaced by more intensive production systems and monocultures, artificial feeding is replacing or supplementing the use of natural foods and waste products.

The main advances in fish nutrition have occurred for high-value species such as salmon and shrimp through the following:

- A greater understanding of the nutritional physiology and biochemistry of the different species, leading to improved feed composition, including reduction in the use of fish meal (although increased use of fish oils) with the addition of minerals and amino acids of vegetable origin
- Improvement in feed pelleting technology so that feed pellets sink slowly, allowing almost all food to be ingested by caged fish
- The development of larval feeds, such as rotifers and brine shrimp
- The search for substitutes for high-cost feed ingredients

These advances have been complemented by fish breeding programs that select for fish with superior metabolism and digestion. For example, the feed conversion ratio for some Norwegian salmon farms is now below 1:1; nitrogen loading has decreased from 180 kg to approximately 30 kg nitrogen per ton of fish; and less than 200 kg solid waste and less than 5 kg phosphate per ton of fish is produced. The protein retention has doubled and feed costs have declined by about 25 percent since 1990 (Cho et al. 1994; Hardy and Gatlin 2002; Myrseth in press).

Fish to Feed Fish

The use of fish to feed other fish is often considered wasteful and led to the argument that some forms of aquaculture do not contribute to net fish supply (Naylor et al. 1998; Allan 2004). A contrarian view suggests that, based on energy flow values, aquaculture represents a significant ecological advantage over the performance of wild fish (Forster 1999). The feed fish are usually small pelagic fish or so-called trash fish (Kelleher 2005)—that is, fish caught but unsuitable for direct human consumption. Approximately one-third (30 million tons) of the annual global marine catch is converted to about 6 million tons of fish meal and 1 million ton of fish oils, values that have remained relatively stable since the mid-1980s and that are unlikely to increase. Some 52 percent of world fish meal and 82 percent of fish oil supply was used by aquaculture in 2004 (Shepherd 2005; Tacon 2006)—proportions that are steadily increasing (Tacon 2003). Fish oil rather than fish meal is the limiting factor, particularly for carnivorous species, although its use in the diet of noncarnivorous fish is also increasing (Pauly et al. 2001). As a result, prices of fish meal and oil are soaring; this, in turn, is driving the search for vegetable substitutes. Substitutes exist, but the costs of their production and extraction remain high and there is resistance to use of GMO feeds. Fish oil production, or the availability of cost-effective substitutes, is becoming a constraint to the continued expansion of aquaculture. Global supply of compound aquafeed is on the order of 22 million tons, or less than 5 percent of global animal feed supply.

Trash Fish and Animal Wastes

Recent estimates indicate that, in addition to the small pelagic fish used for fish meal production, between 5 and 7 million tons of trash fish are used as raw material for fish meal or are fed directly to fish, particularly in Asia (Tacon 2006). It is clear that the market for trash fish is driving overfishing, particularly by trawlers. Asia-Pacific Fishery Commission (APFIC) (see http://www .apfic.org) has begun to address the complex fisheries management issues involved, which lie outside the scope of this study. The global supply of terrestrial animal wastes is between 15 and 30 million tons, but the use of animal wastes as feed is constrained by biosecurity considerations and veterinary advice.

Relative Efficiency

Fish are more efficient protein producers than livestock and the nutritional value of fish is greater. The conversion ratio for high-energy feeds on modern salmon farms is approaching 1:1, compared with 1.8:1 for poultry, the most efficient livestock converter. There are several reasons for this. Fish are cold blooded, use less energy to perform vital functions, and do not require the heavy bone structure and energy to move on land. Fish catabolism and reproduction are also more energy efficient. Livestock use 54 percent of global fish meal supplies, and this superior feed conversion efficiency of fish is rarely noted by those who argue that using fish meal to feed fish is wasteful.

Sourcing of Feeds in Rural Areas

The supply of fish feeds has been a major constraint to rural aquaculture in Sub-Saharan Africa and elsewhere. Aquaculture production often does not attain the critical mass to warrant fish feed production by feed mills, whereas extensive farming may not yield adequate returns and available wastes may not offer the nutrition required. Clustering of fish farms, organization of farmers, and greater integration of ponds with waste management are among the approaches advocated. In view of the poverty link, public support is justified to pursue the following:

- Screen and map locally available waste products for their feed potential and develop simple and cost-effective methods of increasing their nutritional value
- Support the adaptation of low-cost processing machinery and improve methods of processing and storing farm-made aqua-feeds
- Support the setting standards and monitoring of aqua-feeds for the presence of GMO ingredients and contaminants such as mercury or PCBs, and for provision of advice and monitoring of feed additives such as probiotics
- Review incentive systems for artificial feed production, such as the removal of tariffs on the import of key ingredients for mills producing fish feeds
- Support the introduction, where land availability permits, of integrated production systems using wastes and natural feeds

Supplying Improved Seeds

There is a strong case for both public and private investment in genetic improvement and maintenance of genetically vigorous broodstock. The cost-benefit ratio for the Norwegian Salmon National Breeding Program is estimated to be 1:15, while the pro-poor GIFT program had an estimated economic internal rate of return of 70 percent.

Genetic Improvement of Cultured Species

Genetic improvement of cultured species can significantly increase the aquaculture productivity, contribute to food security, and reduce the impact on the environment. Genetic improvement reduces diseases and requirements for feed, land, and water. The productivity gains result in increased supplies, reduced prices, and improved product quality. Today, only about 1 percent of aquaculture production is based on genetically improved fish and shellfish (Gjedrem 2000), but recent experience suggests that it may be possible to obtain increases in aquaculture that are equal to or greater than those obtained in agriculture (Gjedrem 1997) (table 3.2). Genetic improvement of a few cultured species has led to significant increases in productivity, resistance to dis-

Table 3.2   Responses to Selection for Growth Rate

Species Group	Percent Gain in Growth Rate per Generation	Number of Generations
Salmonids	10.0–14.4	1–6
Channel catfish	12.0–20.0	1–3
Tilapia	12.0–15.0	5–12
Carp	30	2
Shrimp	4.4–10.7	1
Bivalves (oysters, clams, scallops)	9.0–20.0	1–4

Source: Gjedrem 2006.

ease (Argue et al. 2002), lower market prices (tables 3.1 and 3.2), and other desirable attributes.

Other major cultured species (Chinese and Indian carps and the giant tiger shrimp) have received relatively limited attention, and a few species (Yesso scallop, blue mussel, white Amur bream, and milkfish) apparently have not been genetically improved at all (Hulata 2001). However, valuable information about the genetic structure of wild and captive populations of Chinese and Indian carps is now becoming available and the applications of genetics to these cyprinids should increase greatly in the coming decades (Penamn 2005). Most of the genetically improved strains used today were developed through traditional selective breeding (selection, cross-breeding, and hybridization). Public concern and consumer resistance has restricted the use of GMOs, although they hold much promise for productivity gains (Hulata 2001).

Seed Supply and Quality

Timely and adequate supplies of quality seed has been a precondition, in all regions, for scaling up production and adopting aquaculture by new entrants. Networks of private producers and traders dominate the supply of seed to farmers in Asia and are important promoters of production, although poor-quality seed, caused by poor genetic management of breeders and accidental hybridization, is a common emerging constraint. The poor-quality seed undermines the livelihoods of poor farmers and the integrity of the production chain and entire aquaculture economy. For example, in the Yangtze River basin (China), inbreeding has reduced the growth rate of farmed Chinese carps by 20–30 percent compared with wild counterparts, causing a 15 percent loss in revenue (Freshwater Fisheries Research Centre 2001). In addition, inbred fish pose a genetic risk to wild counterparts if they escape to rivers in large numbers and may result in the loss of pure stocks. In some areas, cross-breeding of silver and bighead carps that have a similar appearance has eliminated pure stocks; as a result, the efficiency and capacity of silver carp to feed on phytoplankton have plummeted, reducing growth and compromising the value of polyculture in situations in which the different carp species occupy different pond niches—some feeding on phytoplankton, others on zooplankton. Lack of adequate seed supplies is also a major constraint to aquaculture development in Sub-Saharan Africa.

Dealing with Risks

The risks to the genetic integrity of populations from artificial propagation and genetic manipulation of aquatic species are considerable (Kapuscinski 2005; Nash 2005). There is a major difference between aquaculture and live-stock concerning the scale and extent of genetic interactions between cultured and wild populations: (1) it is virtually impossible to prevent escapes; (2) restocking programs may draw on a limited gene pool; (3) fecundity of aquatic species is far higher than terrestrial species; (4) fertilized eggs and juveniles become widely dispersed in the water bodies; and (5) mass rearing of juveniles under artificial conditions may mask traits, causing juvenile mortality in the wild. Public oversight is therefore essential.

Fish Seeds as Public Goods

The pubic sector has an important role to play in ensuring seed quality and supply mostly by encouraging private sector investment through various instruments (see box 3.1). Public actions, in view of the social and environmental externalities and the infant industry argument, may include the following:

- Producing and distributing seed as a temporary measure until there is adequate demand to drive private investment
- Providing fiscal incentives to encourage investment in hatcheries
- In cases in which broodstock development and maintenance is not commercially attractive, as in many countries of Asia, developing, maintaining, and distributing quality broodstock or fertilized eggs to private sector seed suppliers
- Ensuring adherence to precautionary approaches, codes, guidelines, and best practices, in particular to avoid loss of genetic biodiversity
- Encouraging public investment in gene banks for threatened wild populations
- Encouraging and supporting private research on genetic improvement and measures to mitigate loss of genetic biodiversity

Sharing Scarce Resources

International collaboration and public support can improve the quality of seeds and maintain supplies of quality seeds in developing countries. Low-value/low-input species (a mainstay of food fish supplies) are a priority target

China: The government has responded to seed quality problems by (1) encouraging investment in hatcheries; (2) instituting seed quality control policy measures to improve seed quality management, including the establishment of fish seed certification methods and standards; and (3) encouraging and supporting, by law, the production and distribution of quality seed (Hishamunda and Subasinghe 2003).

Vietnam: The Ministry of Fisheries initiated an Aquaculture Seed Development Program in 1999 to promote the expansion of fish hatcheries and to expand and improve the production of quality fish seed from commercial hatcheries. National Broodstock Centers produce and distribute quality broodstock, provide training in good broodstock management practices and advanced hatchery technology, and maintain gene banks (Mair and Tuan 2002).

National breeding programs for genetic improvement of wild stocks for improved aquaculture production include the National Breeding Program for Atlantic salmon and rainbow trout in Norway (started in 1975) and the Philippines' National Tilapia Breeding Program. Favorable investment conditions have fostered private programs in Chile, the United States, and elsewhere (Olesen et al. 2003).

for public intervention and genetic enhancement programs. Networks such as the International Network on Genetics in Aquaculture (INGA) are needed to circumvent the scarcity of human and infrastructure resources in developing countries; to optimize the use of scarce funds; to build capacity and share experiences; and to bridge a growing north-south "molecular divide."

Disease Management

Disease is an important constraint to sustainable aquaculture, affecting investor confidence, profitability, and trade (Subasinghe and Phillips 2002). Pandemics have devastated shrimp aquaculture, resulting in large areas of abandoned farms and the destruction of local economies. Aquatic animal health is a global issue: diseases spread through trade and transboundary movements, and disease control requires international collaboration, while the solutions (best farm practices, treatments, and vaccines) need technology transfer, capacity building, and trade. Disease outbreaks in cultured fish also pose a threat to wild fish populations.

A wide range of factors contribute to the globalization of fish disease problems: increased trade, including in ornamental fish; intensification of fish farming; and increasing movement of broodstock and juveniles. Other factors include the introduction of new species for aquaculture development; enhancement of marine and coastal areas through the stocking of aquatic animals raised in hatcheries; unanticipated interactions between cultured and wild populations of aquatic animals; poor or lack of effective biosecurity measures; and lack of awareness on emerging diseases (Bondad-Reantaso et al. 2005).

The World Bank estimated that in 1997 global losses from aquaculture disease was on the order of $3 billion per year (cited in Subasinghe, Bondad-Reantaso, and McGladdery 2001). Losses caused by shrimp diseases were estimated at $3 billion for 11 countries for the period 1987–94 (Bondad-Reantaso et al. 2005). Annual losses for all species in China are estimated at more than $120 million (Shilu and Linhau 1999). Sea-lice mortalities have cost the EU salmon farming industry an estimated €14 million per year, while viral disease (infectious salmon anemia [ISA]) cost the industries of Norway, Scotland, and Canada on the order of $60 million (MacAlister, Elliott, and Partners 1999).

Threats to Wild Populations

As already noted, disease can be transmitted to wild fish through the water, by escaped fish, or by enhancement programs using diseased seed (Nash 2005; Walker 2004). Exotic pathogens and parasites can be introduced by the unregulated movement of live aquatic animals, while fish farms can amplify and retransmit diseases and parasites occurring in the indigenous wild populations. Among the well-known examples are sea lice in Atlantic salmon and viruses in shrimp (Goldburg, Eliot, and Naylor 2001; Subasinghe, Bondad-Reantaso, and McGladdery 2001). The risk of disease transfer from farmed fish to wild populations is difficult to quantify. Even disease- or parasite-resistant fish, although without disease symptoms, might act as reservoirs, transmitting pathogens to wild fish that are less protected from the particular pathogen.

The Industry Response

Serious disease outbreaks have been catalysts for profound changes in the structure and operation of the aquaculture industry and have led governments to address more seriously the issue of aquatic animal disease management. There has been an increasing reliance on trade restrictions, often implemented without reference to the internationally agreed procedures and protocols set out in the SPS agreement of the WTO that are intended to promote free trade (Chilaud 1996; Walker 2004).

Reducing the Risk

Reducing the incidence of diseases significantly improves production efficiency and profitability. It also reduces the use of chemicals and antimicrobials and minimizes the impact on the environment and wild stocks. A range of BMPs provide practical operational guidelines to prevent diseases and can be incorporated as mandatory conditions for licensing. These BMPs include the use of certified disease-free stocks, maintenance of sensible stock densities, stress management methods, good water quality, proper nutrition, proper storage and use of feeds, and guidance on the use of antimicrobials. A range of improved technical approaches support these BMP, including the following:

- Improved methods for rapid disease diagnosis and pathogen detection
- Closed-cycle breeding and selection for disease resistance
- Use of vaccines as successfully used in marine finfish in Japan and in Norway where vaccination has greatly reduced the use of antibiotics for Atlantic salmon
- Selective breeding programs have produced disease resistant carp (to dropsy), salmon (to vibrio), trout (to furunculosis), and catfish (to catfish virus) (Dunham et al. 2001)
- Approaches to assess the ecological carrying capacity of the aquaculture system (ICES 2005)

The public role parallels that of the livestock industry, in particular the development and implementation of a national strategy designating clear responsibility for aquatic animal health (Bondad-Reantaso 2004). The strategy includes provisions, among other things, for the following:

- Diagnostics, health certification, and measures to ensure regular monitoring and reporting of fish health by farmers or veterinary services
- Mechanisms to identify disease outbreaks and coordinate responses, including quarantine, alerts, disease zoning, and international collaboration
- Capacity building among veterinary professionals, fish farmers, and producers' associations in fish health care
- Enforcement of regulations on transboundary movement of live aquatic animals
- Application of approaches, such as risk analysis, as a basis for policy development and decision making, which are increasingly required by importing countries
- Creation of an enabling environment for the development of vaccines and disease-free strains

Developing and Implementing National Strategies

Existing codes and guidelines such as those developed by OIE, FAO, International Council for the Exploration of the Seas (ICES), NACA, and others (see annex 2) can serve as the basis for national strategies and plans for aquatic animal health. Regional cooperation is a cost-effective means of building national strategies and strengthening national capacities for aquaculture health management (Subasinghe, Arthur, and Shariff 1996). Regional reference laboratories and centers of expertise can provide specialized diagnostic services, advice, and training; facilitate standardization and validation; and underpin collaborative research on diseases. International development assistance and cooperation can play an important role in helping developing countries to develop and implement national aquatic animal health strategies and plans. For example, cooperation among FAO, OIE, and NACA helped establish a quarterly aquatic animal disease reporting system, which has been in place since 1998.

Competing for Freshwater

Freshwater aquaculture accounts for 57 percent of global aquaculture production, and an adequate supply of quality freshwater is a primary requirement. Aquaculture is an efficient user of water. Fish live in but do not necessarily consume water—the water is used merely as a medium for waste removal and oxygen supply. Most forms of intensive aquaculture have much higher water productivity than most crops (Rijsberman 2000), and recirculating aquaculture systems used for eel and other species consume negligible quantities of water (see annex 6). Nevertheless, there is need not only to increase water productivity in aquaculture, but also to see aquaculture as a cost-effective means of conserving, recycling, and buffering water supplies at farm and community levels. Strategies for sustaining an adequate water supply for the expansion of aquaculture may include the following:

- Improving the competitive access of the sector to water resources—that is, the recognition of aquaculture by government as a legitimate claimant of resources and its assimilation into natural resource allocation through integrated management of catchments and coastal areas
- Increasing awareness on the water productivity of aquaculture and the potential for its integration into irrigation systems and agriculture farming systems
- Reducing water demand (and the discharge volume) for aquaculture by scaling up the use of water-efficient intensive aquaculture systems and supporting innovation of new water-efficient aquaculture systems
- Enforcing regulatory measures to reduce pollution by other users to increase the availability of good quality water
- Promoting the adoption of water-related BMPs

TECHNOLOGY TRANSFER AND CAPACITY BUILDING

The section highlights elements of the Asian experience in technology transfer and capacity building, the main lessons learned from the experience, and their relevance to other regions (see annex 4, The Regional Framework for Science and Technology Transfer in Asia, for details of the institutions involved). While some distinction can be made between technology transfer and its diffusion, the two processes are regarded as a continuum, the diffusion as a process of extension to capitalize on the benefits of the technology transfer (Konde 2006; Simpson 2006). Critical issues in technology transfer include the lack of capacity of recipients and inequalities among recipients to access and apply technology (for example, between large agribusiness and small farmers).

The flow of aquaculture science and technology in Asia followed several paths: (1) within the region from countries with advanced aquaculture to other countries in the region; (2) from other more mature sectors like crop and livestock husbandry; and (3) into the region from countries outside the region with advanced disciplines and technologies relevant to aquaculture. During this process, the human and institutional resource capacity was strengthened at many levels from policies, regulations, and planning to financial services and specific technical disciplines. A wide variety of organizations, agencies, and institutions were involved, including the private sector, regional indigenous organizations, development banks, national governments and institutions, and bilateral and multilateral assistance agencies.

Technology transfer was facilitated by and built on a long-standing base of traditional production systems, markets, and education systems, and was hampered by a highly fragmented production structure (a dispersed multitude of small farmers), weak extension services, and initially, lack of documentation or understanding of traditional production systems. With rare exceptions, aquaculture technology transfer in Asia differed little from the experience in agriculture and animal husbandry. The most significant and lasting impacts of this process were the development of national and regional institutional frameworks and the human resource base, and the expansion and consolidation of the scientific foundation of aquaculture.

Drivers of Technology Transfer and Capacity Building

Key drivers of aquaculture development and attendant technology transfer and capacity building in the region included (1) market forces, (2) investments by government and in particular by the private sector, (3) sustainability issues, (4) initiatives of donors and assistance agencies, (5) regional coordination, (6) novel approaches to extension, and (7) intersectoral catalysis. Some of these drivers and their roles in technology transfer are highlighted below, others are discussed in more detail in annex 4, The Regional Framework for Science and Technology Transfer in Asia.

Regional Coordination

Effective cooperation among countries of the region has played an important role in facilitating the identification, sharing, adoption, and extension of technologies. This is partly due to the institutional frameworks established in the early 1980s, from which a number of lessons can be learned (De Silva 2001). A regional strategy on aquaculture development, agreed to by governments and reflecting their common priorities, provided coherent guidelines and focus for various initiatives at the international, regional, national, and local levels. The continuity of the initiatives was underpinned by government ownership, which also ensured their uptake into national or state government policies and programs. Bilateral assistance to individual countries benefited from being harmonized with a regional strategy, receiving inputs of expertise and experiences from other participants (or members of the regional community), and having the results of such assistance shared more widely in the region. Bolstered by government commitments to regional cooperation, the intergovernmental network (NACA) has welded what would otherwise have been isolated, disparate, and diffused national and institutional activities and external assistance to national efforts. NACA's history and development coincides with the stages of development of Asian aquaculture.

Overcoming Weaknesses in Extension

Weak extension services have hampered a more effective diffusion of technology, particularly to small farmers. The traditional government-based research and extension system is not sufficiently responsive to the new challenges and opportunities, neither of aquaculture technologies and markets nor to the demand of farmers. Alternative approaches have emerged to address this problem, such as the following: (1) collaborative links in the public sector, such as pooling of public resources under one agency in Sri Lanka; (2) public-private partnerships, such as government-NGO extension services in Bangladesh, contract farming in Indonesia, and extension services by input (fish feed) providers; (3) One-stop Aqua Shops (OASs), service centers for farmers and fishers who are interested in aquaculture (see box 3.2); and (4) the organization of farmers into producers' associations. This organization has empowered small farmers to effectively demand and benefit from technical assistance and services and has facilitated the government task of providing cost-effective services. For example, India's Marine Products Export Development Authority collaborates with a group of institutions. Farmer associations facilitated the promotion of self-regulating or voluntary management mechanisms to complement regulations and to adopt BMPs. This accelerated the transfer of technology and capacity needed for their application. The feedback from NGOs and farmers groups to research and development (R&D) and policy increased the relevance of policy, research, and technology (Yap 2004).

Intersectoral catalysis. Aquaculture also benefited from well-developed infrastructure and technical expertise in many disciplines (for example, genetics and breeding, engineering, biotechnology, feed and nutrition, disease control, and health management). Similarly, aquaculture has built on education,

As a result of demand from fish farmers, the One-Stop Aqua Shop rural development experiment was started in India by NACA through its STREAM Initiative, and the scheme has spread to Pakistan and Vietnam. OAS provides the following benefits:

- Serves as local contact points for rural banks, aquaculture suppliers, and fisheries departments and helps to introduce aquaculture through self-help groups
- Provides information about supplies and prices; information about fisheries department schemes; and advice, information, and application forms for microcredit from rural banks
- Assists in establishing links with villagers
- Provides information about aquaculture activities of different countries and how this information can best be used locally
- Organizes exposure visits and training programs for interested people who want to do aquaculture
- Provides fisheries materials to people, including fishing nets, pitchers, fingerlings, and feed

Source: STREAM Initiative. Accessed 2006 (available at: http://www.streaminitiative .org/).

extension, and communications systems; existing feed manufacture, market, product handling, and transport facilities set up for livestock and fisheries; and proven protocols developed for other sectors, particularly livestock (for example, in epidemiology, risk assessment and management, early warning and preparedness for disease outbreaks, and quarantine and certification in health management). The existing capability for the development of drugs, prophylactics, antibiotics, and vaccines for livestock was readily retargeted to aquaculture. National agriculture research systems, built in collaboration with the CGIAR centers and through the support of donor agencies and the Bank, also facilitated research priority setting and planning. Thus, aquaculture knowledge and technology diffused across sectors in Asia. In terms of an aquaculture development strategy, the lesson is to adapt successful models in other sectors for aquaculture.

Various reviews made of external assistance programs to aquaculture in the Asian region indicate a number of guiding principles that are not unique to aquaculture (see box 3.3). By the time external assistance to Asian aquaculture intensified in the 1980s and 1990s, governments and donors had the advantage of adapting the lessons learned from the more mature economic sectors.

Following are guiding principles from reviews of external assistance to Asian aquaculture:

- External assistance is applied as a catalyst to national initiatives to build their own capacities, not as a substitute for what the national governments lacked in resources and capacities.
- Relevance of a project is ensured with a broad ownership and participation of governments and national stakeholders (in multilateral programs), or government and the primary stakeholders (in bilateral programs) in its planning and implementation.
- Sustainability and continuity of projects are reinforced by capacity building of regional and national personnel and institutions.
- Capacity building of institutions and organizations is more cost-effective when a project builds on existing capacities in the region or country rather than tries to establish or develop a parallel capacity.
- Multilateral and multi-institutional collaborative projects can be designed and coordinated to enable the various partners to add value to each other's efforts and results, and can mutually strengthen capacities among partner organizations.
- A regional program that reflects the common priorities of participating governments rather than the overriding interest of one or two participants serves all parties better and draws sustained commitment and interest from all the participants to the program.
- Design and objectives of a program or project should not make it vulnerable to being hijacked by an interest group.

Source: FAO/NACA/SEAFDEC/MRC/WFC 2002.

Lessons Learned

The lessons learned from the Asian experience in the transfer of technology are largely not specific to aquaculture and mimic the experience in agriculture. The more important conclusion includes the following:

- Despite its ancient roots in the Asia region, aquaculture, as an emerging industry, has had to negotiate the range of hurdles and issues faced by any sector that seeks to establish itself as a recognized economic sector with legitimate claims to resources and government attention.
- The success of various modes of technology transfer is contingent on the capacity of recipients to promote, adapt, adopt, diffuse, and implement technological innovations. Therefore, technical development cannot achieve last-

ing effect unless due consideration is given to the supporting institutional architecture. In Asia, the early development of institutions, together with subsequent technological advances, is credited with the successful and rapid growth of the industry.

- The most effective mechanisms for the transfer and diffusion of technology in the region have so far been as follows: (1) national will and commitment and development strategies with a long-term view combined with appropriate institutional arrangements; (2) regional intergovernmental indigenous organizations, particularly NACA and FAO's Technical Cooperation between Developing Countries Programme; (3) recent innovative extension methods, establishment of producers' associations, and emerging contract farming; and (4) long-term regional and inter-regional programs like FAO's Aquaculture Development and Coordination Programme (ADCP) and the Asian Institute of Technology (AIT) Outreach Program. Private sector joint venture enterprises, although successful, are less effective in the diffusion of technology.

- Diffusion of transferred technologies has been a problem. As in agriculture, the slow transfer and adoption of modern science and technology which is a basis for knowledge-intensive aquaculture constitute a critical bottleneck. The fragmented nature of the production sector defeats traditional extension systems and makes it difficult to expand the use of new technologies. There is need to document and scale up successful alternative approaches, pilot new approaches, and create partnerships among technology-generating institutions, private sector industrial/marketing concerns, and clusters of small farmers, especially for the purpose of increasing rural incomes.

- The establishment of regional intergovernmental organizations is a long-term phased process and requires sustained support, before and after formalization, until the organization attains functional stability. This has been the experience with NACA and other regional organizations in Asia and elsewhere.

- Lack of interdonor coordination and multiple projects caused wasteful overlap and taxed the resources of host developing countries. Partnerships and coordination among donors would render the external assistance more effective. Links between donor partnerships and NACA-type network organizations would increase efficiency and strengthen these regional bodies and their national lead centers. Such partnerships seem to be on the increase among donors.

- Aquaculture draws on the expertise, technology, and infrastructure of other related and mature sectors. Therefore, the state of development of these sectors should receive special attention in development planning.

- The aquaculture experience in technology transfer and capacity development repeats the experience of agriculture. As aquaculture shares the use of land and water with agriculture, it presents a good case for intersectoral collaboration at the national and institutional levels with a focus on increased productivity, use of limited resources, and poverty reduction.

Possible Lessons for Africa and Latin America

On the basis of the Asian experience, the establishment of an international collaborative mechanism offers many advantages. This need not necessarily be a clone of NACA,[11] but it would possess the same core attribute of enabling collective commitment and action with extreme cost-effectiveness. A broad-based regional program on one or more carefully selected priority topics of regional importance could be a catalyst to closer government, private sector, industry, and NGO participation; stimulate technically efficient and responsible production and processing; and draw interest and assistance from developed country trading partners. Because of common regional interests, this could be the core program or launching platform of any regional network organization.[12]

The establishment of an interregional Technical Cooperation among Developing Countries (TCDC) fund would considerably strengthen and facilitate the current TCDC activities among African and Latin American aquaculture institutions (and link to Asian institutions) through an exchange of expertise and training of personnel, and could complement small-farmer to small-farmer exchanges in the region and to other regions. This form of collaboration could extend to the establishment of joint venture enterprises, possibly reducing investor risks by accessing investment insurance facilities.

Lack of access to all types of information is a serious bottleneck in Africa and to small and subsistence farmers everywhere. A mechanism to collate, screen, and disseminate proven and appropriate technology is a priority. The OASs, based on Internet technology when possible, is a possible model that could facilitate the flow of knowledge and diffusion of technology at regional, national, and local levels, especially for new entrants. The feedback from a network of such shops could also serve to inform policy decisions, market and trade initiatives, extension strategies, and training and education content.

Although the focus of this section has been on the transfer and dissemination of technology in Asia, many key innovations occurred in Norway and the United States. Catalytic breakthroughs in breeding, nutrition, vaccines, and disease-free strains of shrimp all originated in these countries and were subsequently transferred to Asia and elsewhere. A further range of science-based management measures have also become part of industry practice in developed countries. Many of these softer innovations still need to be suitably adapted and mainstreamed into practice in developing countries.

In summary, modern aquaculture is knowledge based, moving rapidly toward intensification and productivity gains and countering resource constraints with knowledge-based advances. With a focus on proven fish culture systems, developing countries can harness this knowledge in a cost-effective manner through regional networks, through both south-south and north-south cooperation, and by establishing a favorable investment climate for FDI.

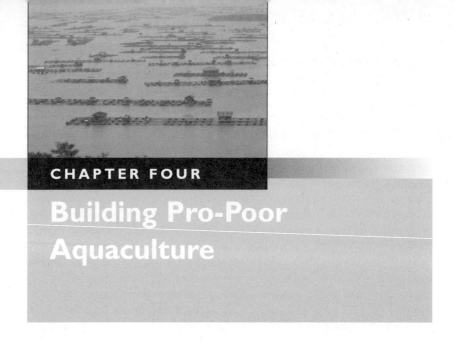

CHAPTER FOUR

Building Pro-Poor Aquaculture

The purpose of this section is to assess the potential of aquaculture in contributing to several of the Millennium Development Goals (MDGs) and is based mainly on the issues and lessons emerging from Asian experiences. It examines the objectives, approaches, rationale, and challenges for pro-poor aquaculture. Aquaculture contributes to the MDGs on the reduction of poverty and hunger by providing income and essential nutrients to combat malnutrition. Aquaculture can also contribute to the empowerment of women, and its contribution to human health and environmental sustainability has already been described. The capital and knowledge character of the sector can be a barrier to entry for the poor, and the structural changes taking place in the sector, in particular in East Asia, can cause small producers to be crowded out by large commercial enterprises. This section highlights the need for sound public policies and strategies to help the weaker segments of society adapt to these changes and take advantage of the new opportunities.

IMPACTS OF AQUACULTURE ON POVERTY AND LIVELIHOODS

Incomes and Employment

Aquaculture generates income for the rural poor through direct sales of aquaculture products and employment in fish production and services and especially in processing. In southeast Asian countries, for example, fish farmers generally earn higher household incomes than other farmers (see box 4.1). In China, aquaculture has been an integral part of rural strategy, absorbing surplus rural labor that was released following agricultural reform. More than 3 million Chinese have found employment in aquaculture since 1974 (see figure A4.1).

Food Security

In many countries, the average market price of fish is lower than that of meat and poultry. The low product prices can make cultured fish highly accessible to even the poorest segments of the population. For example, when the price of meat increased in West Africa in the late 1990s because of the devaluation of the CFA currency, many poor households shifted to dried fish. In landlocked countries, such as Nepal and Laos, the poor largely depend on freshwater aquaculture for their animal protein intake. Furthermore, aquaculture production may reduce fish prices, increasing access to fish by poor households.

Women in Aquaculture

In Bangladesh and Vietnam, more than 50 percent of workers in fish depots and processing plants are women, and although salaries of these workers are still quite low ($1–$3 per day), they are significantly higher than wages earned from agricultural activities. Shrimp seed collection and fish marketing are important sources of employment for rural women. Aquaculture in the Mekong delta has contributed to a decrease in urban migration by young women and prevented women from being forced into prostitution, reducing the risks of spreading HIV/AIDS. Projects that targeted women and poor households have provided access to land, water, credit, and extension that otherwise would not be affordable; however, in a number of cases, the poor lost control of the land and water resources after withdrawal of project support.[13]

LESSONS FROM ASIA

The development of Asian aquaculture, with its major structural changes driven by trade and technology, provides many lessons on implementing pro-poor aquaculture. Aquaculture has taken three distinct development pathways that have merged and overlapped as social and economic conditions have changed (see table 4.1).

- **Static model.** The static model centers on the poorer fish farmer who makes a modest investment and reaps a significant benefit, but a benefit that is often insufficient to lift the household out of the poverty trap. For example, in a World Bank project in Bangladesh, 16,000 of the poorest households were targeted. Although livelihoods improved, the increment was considered inadequate to prevent a slide into poverty. This model can improve livelihoods and open opportunities to graduate to the transition pathway through modest and gradual investment leading to diversified farming systems, improved cash flow, and net household income. Nevertheless, experience shows that many farmers do not transition—implying that the sector is largely driven by dynamic and profit-oriented producers.
- **Transition pathway.** The transition pathway depicts the more advantaged farmer or small enterprise. Access to knowledge, markets, and capital underpins investment and a scale of production that generates a significant surplus and offers a way out of poverty for the household.
- **Consolidation pathway.** The consolidation pathway embraces both corporate and community models of industrial agribusiness. Corporate investment, often in vertically integrated farms, generates employment along the value chain from feed supply to processing plants. A community of well-organized small farmers can benefit from economies of scale in joint activities, while retaining the flexibility to adapt to change. For example, in Thailand, the average area of shrimp farms is 2.7 ha, while in the Philippines the vast majority of brackish-water fishponds are more than 10 ha. In Thailand, farms are substantially more productive, largely intensive, and owner operated; in the Philippines, many farms are semi-intensive and often managed by corporate caretakers.

Table 4.1 Characterization of Aquaculture Development Pathways in Asia

	Static Pathway	Transition Pathway	Consolidation Pathway
Model description	– Small farm/household level innovations and partial modification of production systems – Modest intensification	– Larger farms, progressive farmers, SMEs – Investment in capital and skills to intensify farm production – Loose farmer organizations	– Tightly integrated farmer organizations, corporate model, or both – Substantial capital investment in facilities – Land conversions for large aquaculture production
Outcomes	– Significant income increase and livelihood buffer for small and marginal farmers	– Links farmers to markets and input supplies	– Employment and economic growth along the commodity and input supply chains – Generates employment along value chain
Prospects	– Often not a way out of poverty – Improves livelihoods – Insufficient surplus to invest in education, improve nutrition, and shift household to off-farm incomes	– A way out of poverty	– Supports contract farming and "nucleus estates"

Source: Ahmed 2006.

Aquaculture's contribution to poverty reduction in Asia has evolved along all three pathways. For example, in the Mekong delta, catfish culture gradually evolved from subsistence family-based systems through Vietnam's VAC system to more commercialized agribusiness.

Asian experience shows that success in pro-poor aquaculture requires an enabling environment (box 4.3), which includes the following key elements:

- **Land use and incentive policies.** In China and Vietnam, liberalizing land use policies, in particular, the rezoning of rice land for aquaculture, allowed substantial increases in incomes. By limiting commercial shrimp farm size to 20–50 ha, Indonesia created a basis for nucleus estate shrimp farms. Creating opportunities for the landless to lease unused public water bodies, such as drainage canals, for aquaculture provided an entry point for the poor (see box 4.2). In West Bengal, a shift in economic policy to export-led growth opened the way for what is now a major shrimp culture industry (Chopra and Kumar 2005).

- **Access to knowledge and technology.** Dissemination of knowledge on proven technologies combined with credit has fostered huge growth in productivity. Through adaptive research and extension of improved technologies, pond productivity in China grew from 765 kg/ha in 1980 to 4,900 kg/ha in 2000—an increase of some 640 percent. Similarly, rice-fish culture production rose more than four times by adapting technologies from pond aquaculture (Xiuzhen 2003). In 22 Indian states, Fish Farmers' Development Agencies trained more than 550,000 farmers, improved technologies, introduced carp polyculture in more than 450,000 ha of fishponds, and increased pond production of Indian carps from 50 kg/ha to about 2,200 kg/ha in the 1974–99 period.

- **Integrated farming systems and plans.** Adapting polyculture technology, Chinese, Vietnamese, and Bangladesh rice farmers have extended aquaculture into their rice paddies, increasing fish production by fourfold in less than a decade and earning a net income of about $1,800 per ha per year in China. Technically, this could be emulated in millions of hectares of rice-growing areas of Asia and Africa (Xiuzhen 2003; Halwart and Gupta 2004; Akteruzzaman 2005).

- **Community-based aquaculture.** Community participation in the allocation of leases over public waters is a widespread practice; community guarantees facilitate long-term pond leases (Radheyshyam 2001). Equitable distribution of benefits and strong leadership are among the success factors. For example, in Bangladesh, a combination of group leasing of ponds and micro-credit empowered women in 175 groups by providing income and increased household food security.

- **Organizing producer groups.** This is a classical strategy common to smallholder producers. In Tamil Nadu (India), a shrimp farmers' association has a voluntary code of conduct, it controls the quality of inputs, monitors

Dr. Modadugu Vijay Gupta was awarded the 2005 World Food Prize for the development and dissemination of low-input freshwater fish-farming technologies for more than 1 million poor farmers and families in Asia. This was achieved through research breakthroughs, through their practical application in the field, and by securing broad-based political and civil society support for an international effort. The key elements of this pro-poor aquaculture success story included the following:

- Focusing on the poor, including landless farmers and women, opening opportunities to become fish farmers using readily available resources.
- Researching new aquaculture technologies targeted at the poor. The techniques used low-input, low-cost polyculture systems, recycling at farm level, and a focus on unused water bodies such as derelict ponds, canals, and use of species capable of surviving water shortages or thriving on garden waste.
- Raising political awareness and securing funds and international support for pro-poor aquaculture.
- Developing and securing the involvement of NGOs.
- Initiating capacity-building programs to build a critical mass of scientists, extension workers, and farmers.
- Identifying and coordinating the distribution of better breeds of fish and developing international fish biodiversity protocols and policies, including protocols for the transfer of fish germplasm.

Source: The World Food Prize Laureates (available at: http://www.worldfoodprize.org/Laureates/laureates.htm (accessed March 2006).

ponds on a 24-hour basis, and uses collective bargaining to market products (Kumaran et al. 2003).

- **Innovative institutional arrangements.** Coordination of policies and institutions has removed bureaucratic constraints, for example, by streamlining food safety and export procedures. In India, the OAS extension and credit services provide a suite of support services to rural fish farmers, while an aquaculture module in eChoupal helps farmers secure fair product prices (see annex 4). Corporate approaches, such as those of eChoupal in India and nucleus estate shrimp farming in Indonesia, show that large commercial enterprises can contribute to poverty reduction by supplying leadership, knowledge, and innovation. NGOs have supplemented government programs, for example, by strengthening farmer technical and financial capacity in Bangladesh and Cambodia and by providing microcredit to the rural poor and women. NGOs have provided an independent and watchful eye on equity and environmental issues.

CREATING AND DISTRIBUTING WEALTH THROUGH AQUACULTURE

Aquaculture can have an important role to play in poverty reduction, and the robust growth of aquaculture compared with the relative stagnation of crop agriculture suggests that aquaculture can be a major rural growth sector (Karim et al. 2006). Government support for aquaculture in the 1980s largely assumed that a trickle-down effect would benefit the poor, or simply ignored the pro-poor dimension. In the absence of a coherent pro-poor approach, however, too often aquaculture benefited the upper strata of the society at the expense of the poor, generated negative externalities, created conflicts over rights to land and water, and even evicted poor people from public land or waters on which their livelihoods were based (Hagler 1997; Edwards 2000). Rising knowledge and capital requirements in aquaculture may be major barriers to entry for the poor. Three questions emerge:

- **Can aquaculture help the poorest?** Experience shows that even the landless can find an entry point through rehabilitation of unused sodic lands or lease of public waters. Moreover, those lacking entrepreneurship, motivation, and persistence can benefit through employment creation. While some studies indicate that poorer households may benefit more (in relative terms) than the richer ones (Irz et al. n.d.), other studies suggest that the better-off households tend to be the main beneficiaries of projects targeting the poor (see, for example, Hallman, Lewis, and Begum 2003). In either case, the most disadvantaged groups require sustained and comprehensive support to benefit from aquaculture.
- **Does smallholder aquaculture have a future?** Given the trend toward consolidation and vertical integration, changing markets, and conditions of trade, smallholders face an array of challenges. Only in 2004, following massive industry consolidation, have corporate salmon farms in Norway demonstrated significantly higher returns to labor than smaller owner-operated farms (Fiskeridirektoratet 2004), indicating that the owner-operated farms retained advantages in adaptability over several decades. This and similar experiences in Asia indicate that well-organized and well-informed smallholder producers can thrive, and that the establishment of a critical mass of self-sustaining smallholder aquaculture requires sustained nurturing and public support linked to progressive rural and agricultural development policies.
- **Is low-value food fish production a viable strategy?** Where favorable circumstances exist, this strategy is viable and sustainable. In many areas, it is a mainstay of local economies—more than 40 percent of global food fish production is composed of low-value herbivores, omnivores, and filter-feeding fish. Despite the poor record of subsistence aquaculture in some regions, a compelling policy argument can be made to support the production of low-cost food fish. Low prices infer low returns and extensive culture systems to take advantage of natural foods and wastes, and extensive culture requires capital such as ponds—capital the poor may not have. Under conditions in which land, water, and labor have limited alternative uses, extensive culture requiring modest capital and knowledge (for example, culture of carps and omnivores) can be an attractive pro-poor production model.
- A strong counterargument can also be made for poor producers to grow high-value species, which may require less capital and generate higher

returns—but only in cases in which there is access to the markets for such species and risks are manageable. Thus, coherency between twin objectives—production of cheap food fish and reduction of poverty through rural fish farming—requires an astute balance of policies and incentives. Production of low-value fish may deliver more benefits to the poor consumer than to the poor producer. Gradual intensification, taking full advantage of natural productivity, may help increase supplies, while moderate production costs keep fish prices within the purchasing power of the poor.

As a first step, the IFIs and other donors, working with their partners, can (1) raise awareness of client countries on the value of aquaculture in poverty alleviation and wealth creation, (2) advise on sector diagnostics and policy approaches, (3) illustrate how aquaculture can complement national rural development and environmental strategies, and (4) seek to integrate appropriate pro-poor aquaculture into CASs, PRSPs, and other policy and planning instruments.

Policies and Awareness
Awareness and Mainstreaming

Aquaculture can be mainstreamed into the policy and planning matrix along several complementary axes: growth and employment, equity and poverty alleviation, and food security. There are some good examples in Asia. For example, in the Philippines the latest Medium Term Development Plan (MTDP) and the Fisheries Code targets modernization of aquaculture technology for the poor.[14] In Vietnam, the government's aquaculture development program for 1999–2010 sets out the country's vision to prioritize aquaculture development for reducing hunger and poverty.[15] In Bangladesh, the national PRSP identifies a clear link between aquaculture and rural economic growth.[16] However, aquaculture receives scant recognition in Thailand's policy and planning documents, underlining the fact that mainstreaming of aquaculture remains a "work in progress."

Enabling Policies

Application of pro-poor enabling policies, such as those described previously and in annex 4 (Philippines, China, Vietnam, and Indonesia) require political will. Ideally, the policy instruments will provide a favorable investment climate, establish aquaculture's claim as a legitimate user of resources, and provide for environmental safeguards. A pro-poor policy framework will define the rights and obligations of producers and allocate preferential rights to the poor for aquaculture in public waters (Cullinan and van Houtte 1997). It will facilitate water and land tenure for aquaculture purposes and deliver access to knowledge and technology. An effective framework is likely to establish participatory processes to guide sustainable aquaculture development and serve as a platform for cooperation among public agencies. A comprehensive national aquaculture plan can provide a road map for public and private sectors, financial institutions, and the international community. Clarity in the respective roles of the public and private sectors transmits clear signals to investors, while establishing a basis for cooperation, synergy, and public-private partnerships. The policy framework will build bridges to other sectors so that farmers can benefit from road and infrastructure development, and access finance and domestic and export markets. These policies need to be developed by national decision makers, based on the participation of all national stakeholders. International partners can support this process by brokering knowledge and experiences from other countries and sectors and by supporting training and other facilitating activities.

Coordination and Sustained Public Support
Coordination

Coordination between institutions is fundamental to the delivery of pro-poor policies and, more generally, to the emerging aquaculture industry. Usually a plethora of institutions is involved in land and water lease, environmental control, sanitary measures, pro-poor programs, and trade; responsibilities and jurisdiction may be split between federal/central and state/local authorities. Coordination among agencies is improving—for example, specific spatial regimes have been created for aquaculture in Malaysia (aquaculture investment zones); the Philippines (mariculture parks); and Indonesia (aquaculture zones, targeting nucleus-estate-type export aquaculture). To serve the poor, this coordination must penetrate to the district, village, and farm level—this is where the benefits of such models as the OAS (see box 3.2) are most evident. The international community can support the coordination process by sharing experiences from other countries and supporting investments (for example, for infrastructure or improved spatial regimes).

Sustained Public Sector Support

A key to pro-poor aquaculture is sustained public support with a particular emphasis on measures to ensure that less-advantaged producers have access to land and water, technology, credit, markets, and a fair share of benefits from the production chain. China, Bangladesh, and India have delivered this sustained support for more than two decades, particularly for the following:

- Production and supply of quality seeds where the private sector cannot effectively deliver
- Investment in human capital (Dey et al. 2004) through training and provision of advice to farmers, including operation of demonstration farms
- Extension services to provide knowledge and access to new technologies and to organize producers

- Applied research to develop low-cost culture systems and improved strains with higher growth rates or disease resistance
- Provision of key infrastructure, such as roads or markets
- Regional cooperation for the exchange of information, capacity building, and transfer of technologies

Some of these functions may be most efficiently undertaken through contracts with and incentives to the private sector; through NGOs, producer associations, and specialized agencies; or through strategic alliances between key actors. Public support to aquaculture requires monitoring to ensure that it does not subsidize large capital-intensive operations at the expense of traditional fish farmers (Ahmed 1997, 2004). The international community can support the investments required to produce these public goods, underpinning them with poverty assessments to ensure that they effectively benefit the poor.

Equitable Trade and Poverty Alleviation

Tariffs and nontariff barriers (NTBs) such as excessive sanitary restrictions have affected trade in aquaculture products, both for developed and developing countries. Such trade distortions and disputes (box 4.4) are likely to increase as more cost-effective aquaculture in the developing world captures market share from traditional developed country suppliers.

Tariff Barriers

Tariffs for fish products remain high and are characterized by "tariff peaks" and "tariff escalation"[17] on the more profitable value added products produced by developing countries. Tariffs on fish products in developing countries also remain relatively high, constraining south-south trade. While many developed countries have reduced or eliminated import tariffs, there are indications that countries are using antidumping measures as an excuse for the protection of the domestic industry.[18]

Nontariff Barriers

NTBs, such as technical and sanitary standards, labeling, and traceability requirements to ensure food safety may be deployed to protect domestic producers. The cost of compliance with increasingly stringent food safety regulations also tends to exclude small producers and processors from export markets (Dey et al. 2004). Additional costs can be envisaged in the future as some importing countries move to screen products for bacteria resistance to antimicrobials.

Developing country negotiators at the Doha Round underlined the importance of providing technical assistance and capacity building to developing countries to adjust to WTO rules, implement existing obligations, and fully

Box 4.4   Trade Disputes over Aquaculture Products

U.S. antidumping duties on shrimp imports

In 2003, the U.S. Southern Shrimp Alliance filed a petition to the U.S. authorities alleging that exporters from Brazil, China, Ecuador, India, Thailand, and Vietnam had dumped shrimp on the U.S. market at below-cost prices, triggering a plunge in the value of U.S.-harvested shrimp from $1.25 billion in 2000 to $559 million in 2002. In 2003, shrimp imports from the six countries amounted to $2.67 billion. In 2005, the U.S. authorities imposed antidumping duties of up to 113 percent on imports of certain shrimp and prawns from the above six countries and, although the measures did not break the rising trend in shrimp imports, they negatively affected both volumes and share in imports of the concerned exporters.

Sources: USINFO 2004; FAO 2006; WTO 2006.

U.S.-Vietnam catfish trade dispute

Between January and November 2002, the United States imported 18,300 tons of Vietnamese catfish worth $55.1 million. The Catfish Farmers of America (CFA) complained that Vietnam had captured 20 percent of the $590 million catfish market by selling at prices below the cost of production, and in mid-2003, U.S. authorities ruled that Vietnamese catfish fillets had been "dumped" or sold in the U.S. market at unfairly low prices, resulting in retroactive import duties of 37–64 percent. Catfish import duties were 5 percent before the rulings. Vietnam maintained that its catfish were cheaper because of cheaper labor and feed costs. Subsequently, the U.S. Congress declared that only the native U.S. family, *Ictaluridae*, could be called catfish, effectively preventing the Vietnamese product from being marketed as catfish, and U.S. authorities initiated an antidumping case against Vietnamese catfish. Some half-million Vietnamese live off the catfish trade in the Mekong delta and the catfish dispute threatened the livelihoods of thousands of farmers until alternative markets were found.

Source: Lam 2003.

exercise the rights of membership. Through their roles in global forums, the international partners can advocate for an easing of the burden of tariff escalation in relation to value added fish products, a dismantling of inequitable NTBs to trade, and removal of subsidies. Financial support can include provisions for capacity building to comply with food safety and quality requirements (for example in minimum risk level [MRL] detection) and to establish certification and ecolabeling schemes.

CHAPTER FIVE

Catalyzing Aquaculture in Less-Developed Countries

The focus of this section is Sub-Saharan Africa (SSA) and, to a lesser extent, Latin America. The section examines reasons for weak performance in aquaculture and provides an analysis of prospects for future development. While many of the issues raised are specific to Africa, many are also relevant to the development of aquaculture in Latin America and countries that are nurturing an infant aquaculture industry.

Why Africa? The contribution of the continent to global aquaculture production is negligible and per capita fish consumption is declining in SSA. Sub-Saharan countries, in particular, have considerable untapped potential for aquaculture production. The resources exist—clean water, unused land—but weak institutional frameworks, deficient human capacity, and a volatile investment climate have been barriers to development of sustainable aquaculture. With improving governance, growing urban purchasing power, and increasing recognition of the role of the private sector, the tide may now be turning. Moreover, the World Bank has a clear focus on Africa's development (World Bank 2006). This section examines the constraints and the lessons of past endeavors to launch aquaculture in Africa; it then takes a fresh look at a road map for sustainable aquaculture in SSA. It examines the successes and failures and suggests opportunities for the transfer of technologies, business models, and institutional arrangements from Asia and elsewhere.

To maintain Africa's per capita food fish consumption at present levels (7.8 kg per person-year), supplies should increase from some 6.2 to 9.3 million tons per year in 2020 (Delgado et al. 2003). However, per capita consumption of fish in Africa is currently stagnating and has fallen in SSA. To support these projected future needs, capture fisheries will need to be sustained and, if possible, enhanced; by 2020, aquaculture would have to increase by more than 260 percent (an annual average of more than 8.3 percent) in SSA alone.

There are no physical and technological barriers to a major expansion of sustainable aquaculture in SSA. Many parts of SSA have the basic physical requirements—ample land and water. It is estimated that more than 30 percent of the land area (9.2 million km^2) in SSA is suitable for smallholder fish farming. In theory, if yields (1–2 tons/ha per year) from recent smallholder projects could be replicated, then less than 1 percent of this area would be required to produce 35 percent of the region's increased requirements to 2010 (Kapetsky 1995; Aguilar-Manjarrez and Nath 1998).

Numerous national and regional efforts have been made to launch aquaculture using a wide range of production models and approaches. Numerous projects have demonstrated this potential, but when scaled up, the results have generally been poor. In spite of decades of investment and technical input, aquaculture has failed to thrive where expected, and in many cases remains precarious and marginal. The reasons vary widely from country to country, but several common themes emerge from the portfolio analysis and other reviews. This section explores the lessons from past endeavors to launch aquaculture and takes a fresh look at the road map for sustainable aquaculture in SSA.

THE STATUS OF AQUACULTURE IN AFRICA

Fisheries make vital contributions to food and nutrition security of 200 million people in Africa and provide income and livelihoods for more than 10 million engaged in fish production, processing, and trade. Fish has become a leading export commodity, with an export value on the order of $2.7 billion per year.

Of the 7.88 million tons of fish produced in Africa in 2004, 93 percent originated from capture fisheries. Since the 1990s, recorded output in capture fisheries has stagnated, rising slightly to 7.31 million tons in 2004. In contrast, aquaculture has produced about five times the volume produced a decade earlier. However, at just over 0.5 million tons, this was insignificant in global terms (less than 1 percent) and the contribution to gross domestic product (GDP) is negligible. In a selection of 17 SSA countries, aquaculture contributes on average only 0.9 percent of total animal protein supply (10.1 kg fish and 12.4 kg of meat per capita).

The three top producers were Egypt, alone accounting for 85.6 percent of the total; Nigeria with 6.5 percent; and Madagascar with 1.8 percent. Production has increased, but much more slowly than elsewhere. Only in Egypt has development been notable, from 72,000 tons in 1995 to more than 470,000 tons in 2004, an average annual growth of 26 percent. If Egypt is excluded, the region's compound annual rate of growth from 1990 to 2000 was 17.2 percent, but from an extremely low base. Both growth rates and output levels remain very low and many countries produce negligible quantities. In 2004, the most important Sub-Saharan producers were Nigeria, Madagascar, and South Africa (table A5.8). Sub-Saharan aquaculture exports are dominated by shrimp and abalone, and exports of seaweed, crocodile skins, and ornamental fish have also shown growth. The production systems used in Africa are described in annex 6, Aquaculture Production Systems in Africa.

DIAGNOSIS FOR SUB-SAHARAN AFRICA

Several reviews of aquaculture in SSA concluded that physical potential alone is not sufficient; development investment was largely wasted; and central hatcheries and extension services did not work (Harrison et al. 1994; Moehl, Halwart, and Brummett 2005). Although projects succeeded for short periods, there are few examples of sustainability. Many investment programs have had rather indifferent results, with limited sustainability of production after central support is reduced. There have been few examples of internally generated growth that is typical of profitable and attractive sectors. As a consequence, donors became increasingly skeptical and disaffected with the sector, increasingly convinced that issues of development of aquaculture were more related to markets, policies, and institutions, and that without suitable conditions, investment was likely to be ineffective.

Characteristics of Past Programs

Some common characteristics of these programs included the following:

- An expectation that available water and land resources could, by themselves, lead to natural exploitable potential and create a new option for rural people. The social and institutional contexts in which people engage in aquaculture and issues such as resource access, equity, and policy support received little attention.
- An emphasis on public-sector support usually was linked with the development of aquaculture extension capability within fisheries administrations, many of which have suffered budget cuts and lacked staff salary incentives. The use of state or parastatal agents for broodstock development and hatchery supply had variable and often disappointing results. Intermittent supplies of seeds and feeds also undermined farmer confidence.
- There was an emphasis on small-scale integrated aquaculture, in which freshwater fish farming in ponds is linked with a range of primarily family-supported mixed farming activities. This was coupled with a poor understanding of markets, market margins, and logistics and real returns available to producers. Programs were not market driven but often based on notions of local food security and self-sufficiency. The target groups chosen in early

attempts to foster aquaculture development in SSA often concentrated on subsistence farmers who had little surplus labor or other resources to invest in aquaculture. Participatory approaches to program design were rare.

- There were poor strategic approaches to pooling knowledge in developing seed supply, fertility and feed inputs, environmental aquatic health, and food safety issues, and a limited knowledge of risk issues or appropriate management responses.
- Aquaculture rarely attained the critical mass needed to support segmentation (for example, specialized seed producers) and the rise of service suppliers. Individual farmers were thus often dependent either on weak extension services, or more frequently, on their own efforts for seeds, feeds, and technical and market advice.
- Development banks and suppliers of financial services were not involved in program design and remain largely unfamiliar with the sector, thus constraining private investment.
- Program objectives focused on food supply, supplementary cash income, and integration of ponds into farms rather than on the creation of a commercially viable sector backed by sustainable public and private services.

There was also a failure to learn from earlier mistakes:

few donors have a well-articulated policy for their technical assistance in aquaculture. This is reflected, on occasion, in hasty and uncritical attempts to transfer technology, often not suitable to the needs of the recipient country . . . few analyses were made of the reasons for the collapse . . . to draw lessons for the design of future projects. No analysis has ever been published. (FAO/UNDP/Norway 1987)

Many of the projects launched in Africa replicated or were based on the same premises as projects of colonial times but lacked the supporting analysis to relate the activities and expected results to emerging national programs and priorities. In many cases, personnel from colonial services were employed on development projects in the newly independent nations. They rehabilitated old stations and tried to reestablish what had already failed.

The main exceptions to this pattern are the isolated examples of (usually) large commercial development. These developments include shrimp farms in Madagascar and Mozambique, cage culture in Zimbabwe, and several farms in Ghana, Nigeria, and Malawi.[19] The IFC has financed successful shrimp farms in Madagascar. Aquaculture has been successful in situations in which domestic markets, resources, and available technologies have combined to promote steady and substantial growth, as in Egypt and Nigeria. African countries have repeatedly identified the key constraints as the supply of seeds and feeds and producers' access to technical information.

UNLOCKING THE POTENTIAL

Several of the experiences described above demonstrate possible future approaches in African aquaculture development. Although starting from a low base, some countries have experienced a remarkable growth rate in aquaculture. Uganda has had almost a 3,000 percent increase since 1994 and many others have experienced high growth (table A5.8). Growing urbanization, expanding markets and services, improved skills, opportunities for private sector development, and new production technologies are all playing a role. Productivity gains have contributed by lowering prices and expanding the consumer base. In Egypt, for example, following a drop in the growth rate from more than 60 percent in 1999 to 1 percent in 2001, productivity gains are again driving increased production (6 percent growth in 2003–04).

The NEPAD *Fish for All* Summit has already raised awareness of the potential of aquaculture and aquaculture is featured in many Sub-Saharan PRSPs. At a global level, the Code of Conduct for Responsible Fisheries, the Bangkok Declaration and Strategy for Aquaculture Development Beyond 2000, the Nairobi Declaration on Conservation of Aquatic Biodiversity and Use of Genetically Improved and Alien Species for Aquaculture in Africa 2002, and numerous BMPs provide further guidance regarding policy and practice (see annex 2).

A concerted effort is required if aquaculture is to be mainstreamed into agriculture and rural development plans, into coastal zone management, into industrial planning, and into water resource allocation.

Enabling Conditions

Some of the enabling conditions for sustainable aquaculture in Africa are as follows:

Awareness and Perception

Awareness of aquaculture as a viable commercial undertaking in the public and private sector and among financial institutions improves access to land, water, and financial resources. Recent experiences in Asia and Africa also indicate that farmers adopt aquaculture where certain predisposing conditions are met: (1) perception of the value of fish as food and for generating income; (2) land ownership, ability to rent agriculture land, or in case of the landless poor, secure access to common property resources (water bodies, floodplains, irrigation canals, and/or coastal waters); (3) knowledge of technologies suited to available resources and conditions; (4) a supply of seed; and (5) institutional support, in terms of initial support to new entrant farmers through advice and inputs—for example, extension, seed supply (Van der Mheen 1998; Edwards 2000).

The Declaration adopted by the Heads of State Meeting of the NEPAD *Fish for All* Summit in 2005 called for a range of actions in support of aquaculture, including the following:

- Empowerment of fish-farming communities and stakeholder organizations, including participation in policy making and planning; equitable allocation of resources, particularly for the poor
- Aquaculture to be adequately reflected in the national and regional economic policies, strategies, plans, and investment portfolios, including poverty reduction and food security strategies
- An improved investment climate, including legal and institutional reform and enforcement
- Improved incentives and access to capital for private investors and strategic public sector investments
- Harnessing of the potential and entrepreneurship of small-scale fish farmers
- Fostering of small, medium, and large aquaculture production in a sustainable and environmentally friendly manner compatible with the rational use of land and water resources and evolving market opportunities
- Building of human and institutional capacity with a particular emphasis on transferring technologies and knowledge to small producers
- Conservation of aquatic environments and habitats essential to living aquatic resources and aquatic biodiversity; and measures to prevent or mitigate adverse impacts of aquaculture on the aquatic and coastal environment and communities
- Development of common approaches and positions on international trade in fish and fishery products

Source: Nepad Action Plan for the Development of African Fisheries and Aquaculture. NEPAD 2005 (available at http://www.worldfishcenter.org).

Capture Fishing Industry

The existence of an export fishing industry facilitates market access. For example, the shrimp farms in Madagascar and Mozambique involved investors with links to the fishing industry. Effective sanitary controls and the accompanying legislation are often established to serve the capture fishing industry. Similarly, the use of cold chain infrastructure built to serve the fishing or meat industry can also reduce investment costs.

A Developed Livestock Industry

A developed livestock industry enables spin off of a wide range of skills and experiences, ranging from animal health to effluent control and EIA to the use of feed mills and training institutions.

Urban Markets

Major urban markets, such as Kinshasa and Accra, and those in Nigeria, Egypt, and East African countries offer major opportunities for stable markets (for example, contracts with supermarkets, hospitals, and restaurants). In turn, stable markets allow planned production and create a basis for further investment.

Peri-urban Commerce

The peri-urban zones offer particular opportunities for aquaculture (Rana et al. n.d.). They have easy access to the urban markets or export channels. Wastes generated by other industries, such as breweries or abattoirs, may be recycled. Labor and service industries are available. Access to government services for extension or permits is easier. Infrastructure such as power, water, and transport networks are available. In Uganda, 43 percent of peri-urban fish farmers have a tertiary education (Rana et al. 2005); in the Central African Republic, successful fish farming correlated strongly with proximity to an urban center and level of farmers' education (Kelleher 1974).

Critical Mass

The vertically integrated commercial shrimp farms in Madagascar and Mozambique assume a large part of the risk of the venture. A mature industry has the critical mass to benefits from economies of scale, to spread risk to the service industries and specialist providers. Lower risks means investors can accept lower returns, allowing product price reductions and further market penetration. Building critical mass in an infant aquaculture industry can be accelerated by forging links to the fishing or livestock industry, to farmer or exporter organizations, and to education and scientific institutions. Dispersed subsistence fish farming rarely achieves this critical mass.

Farmer Associations

Strong farmer or fishing industry organizations can provide a platform for fish farmers. Experiences with contract farming, particularly for perishable export crops, may also provide a useful business model, while established rural finance mechanisms can facilitate farmer credit.

National Aquaculture Development Plans

National aquaculture development plans exist in several countries such as Angola, Cameroon, and Malawi. In other countries, aquaculture policies have been formulated and specific legislation enacted. For example, commercial shrimp culture was identified as a target industry in Mozambique's FDI policy for fisheries during the transition to a market economy.

Stable Institutional Arrangements

In one SSA country, institutional responsibility for aquaculture moved ministries five times in three years, dissipating institutional memory and undermining career paths (Hecht 2006). The institutional arrangements for such an industry require stability, a permanent institutional home, or coordinating mechanism. The OASs described in box 3.2 underline this need at the level of farmer operations.

Keys to Sustainable Aquaculture

On the basis of the experiences in SSA, and in Asia, certain key elements in a national aquaculture development program can be identified:

Sector Analysis and Policy Development

A thorough analysis of the sector, its potential, and its constraints forms the basis for a realistic aquaculture policy and strategic plan to be developed through a participatory process. The strategy will define a policy and institutional framework. It is likely to achieve the following:

- Provide clarity on sector objectives (for example, a focus on export-driven commercial aquaculture, subsistence aquaculture directed at social objectives, or some combination of objectives).
- Describe key policy principles.
- Identify the respective roles of the public and private sectors, and the role of and opportunities for FDI (particularly for export products).
- Identify a participatory framework for development, including the means of setting common objectives, coordinating and mutually reinforcing activities, and structuring public-private partnerships.
- Describe the coordination mechanisms in the public sector, including means of resolving jurisdictional overlap.
- Outline a legal and regulatory framework that ensures environmental safeguards without imposing unnecessary restrictions on development, including means of internalizing environmental costs.

- Establish clear principles for resource allocation, including zoning for aquaculture.
- Identify sources of finance for aquaculture—from public and private sources, from internal and external sources—and seek to match those sources with the respective tasks.
- Learn the lessons of the past and focus public interventions on catalytic actions, core regulatory activities, and essential tasks unattractive to the private sector.

The strategy is likely to be implemented through an enabling framework with a number of mutually supportive components.

A Conducive Investment Climate

An investment climate suitable for an infant industry is highly desirable. Good governance, rule of law, and a clear vision of the role of the state all contribute. A favorable investment climate requires a coherent policy framework; attention to issues of tenure; coordination among public bodies and with private stakeholders; and support for science, technology, and capacity building. A suite of public policies such as those introduced in China (see annex 4, National Policy Interventions and Pro-Poor Aquaculture) or Indonesia's regulations that foster a nucleus estate model for shrimp farming provides examples of such environments. Private investment, including FDI, has proven to be a major engine of growth during the startup phase for new aquaculture industries. Special conditions, such as those created for free trade zones, may be appropriate. Incentives may be required to ensure supplies of key inputs. For example, removal of import tariffs on raw materials for fish feeds may provide the incentives feed mills require to manufacture these special rations. Financial institutions, such as development banks, may need government assurances of support during the high-risk startup phase. Similarly, approaches to rural finance already proven in livestock or agriculture may be adapted to aquaculture.

Tenure

In Africa and elsewhere, land and water rights are often complex and poorly defined; lease arrangements are frequently informal and insecure. Aquaculture requires a policy and legal framework that creates clear title (for example, long-term lease) over land and water. Titles need to be protected and freely transferable so that new investors can purchase established fish farms and so that titles can serve as collateral. Excessive bureaucratic process in title emission and transfer creates opportunities for corruption while overlapping jurisdiction creates further hurdles—for example, jurisdiction over mangroves and intertidal zones or over public reservoirs. To provide the incentives to invest in restocking aquatic commons, community or individual rights may need to be created for fish harvesting in lakes and reservoirs.

The Business Models

A Focus on Market-Driven Commercial Aquaculture

Experience shows that successful aquaculture development in Africa and elsewhere has been driven largely by domestic and export market demand. Rising incomes in urban Africa and international demand for shrimp and whitefish, such as tilapia and catfish, create further opportunities.

Poverty Focus

The strategies employed to bootstrap aquaculture through market-driven commercial production must not lose sight of the poverty focus. In other words, commercial aquaculture can provide leadership and build the critical mass necessary to raise the profile of the sector, achieve economies of scale, and create opportunities for the emergence of segmentation and service providers. However, public policy and support is required to ensure that the smaller producers have access to the technologies, markets, and finance for aquaculture. The balance and trade-offs between high-value intensive aquaculture and culture of lower-value herbivorous or omnivorous species will need analysis and informed public scrutiny.

Asia provides a number of lessons. The choice of business model is not necessarily between commercial culture of high-value carnivorous species and small-scale culture of herbivorous or omnivorous species providing food security and household and local consumption. Rather, pro-poor fish culture should be a viable commercial undertaking, whether the product has a social (food security) or commercial objective. Poor people should not necessarily grow low-value fish. Culture of freshwater shrimp in rice paddies; polyculture of higher-value snakehead or catfish with tilapia; and contract farming of shrimp are examples of successful pro-poor business models. As already noted, leasing of public water bodies for restocking or cage culture can provide an entry point for the landless.

Proven Technologies

Use of proven technologies reduces the costs of adaptation and innovation. Public support will be required for pro-poor aquaculture, but it can draw on proven models. Capacity building and institutional development will also require public support. Creation of public private partnerships, leasing of government fish stations, broad stakeholder participation in design of national codes of practice, and BMPs and training programs can reduce costs and make the best use of available expertise and donor support.

Networks

As demonstrated by NACA (see chapter 3, Technology Transfer and Capacity Building), regional networks can have a catalytic action driving dissemination of technologies, building human capacity, reinforcing institutions, and reducing development costs. A network can be a focus for donor interventions and articulate common concerns at global or regional forums. Networks can be malleable, adapting to opportunity, structured at regional or subregional levels, or linked to regional economic commissions for greater political traction.

International Financial Organizations

As shown in chapter 1, The Role of External Assistance and the International Financial Institutions, the IFIs have been of crucial importance—accounting for 92 percent of developing country aquaculture project investment in 1995. The IFIs, including the World Bank, can continue to play a key role through assistance with sector analyses and master plans, building good governance of natural resources, fostering a favorable investment climate, and building aquaculture into country portfolios. The World Bank Group can help reduce investor risk, help align development assistance, and explore partnership opportunities at the regional level. Aquaculture can be a valuable component in a wide range of projects, whether in rural development, livelihood diversification, or land and water management.

CATALYZING SUSTAINABLE AQUACULTURE IN LATIN AMERICA

Key aquaculture developments in the past decade included the spectacular growth of salmon farming in Chile, substantial domestic and foreign investment in shrimp culture and processing in several countries (for example, Ecuador, Brazil, Colombia, and Mexico), regular use of water reservoirs for aquaculture, an evolution to more commercial scale rural aquaculture, and progress in culturing of various native species. Providing less than 3 percent of global aquaculture production, Latin America has vast underused potential. Many Latin American countries have a favorable investment climate and suitable land, water, and environmental conditions. They benefit from proactive producers' associations, relatively low-cost labor, an expanding market for aquaculture products, and availability of raw materials (fish meal and soybeans) for feed production.

The Status of Aquaculture in Latin America and the Caribbean

Total production from aquaculture in the region in 2004 was 1.3 million tons (worth $5 billion), or 6.3 percent of total national fish production.[20] During the same period, output from capture fisheries declined by 18.5 percent, from 24 to 20 million tons. New technologies and production systems have increased output by 235 percent since 1994—an average annual growth rate of 13 percent. However, the combined contribution of aquaculture and fisheries to the region's GDP is modest—about 3 percent. Fish accounts for only about 5 percent of the total animal protein intake; because of a lack of a fish-eating tradition, the higher price of fish, and thriving livestock and poultry industries, consumers generally favor meat over fish (FAO 2005b). Nevertheless, the recent surge of developments in aquaculture has created a greater awareness of the sector's growth potential among investors and planners.

With the notable exception of white leg shrimp, production is based mostly on a small number of introduced species, such as Atlantic salmon and tilapia, and most of the production is from marine and brackish-water culture. Diversification is under way in some countries in an effort to provide a safety net from the boom and bust effects of the shrimp-farming industry. In the past 10 years, this diversification has been evidenced by noticeable increases in the production of other species, such as mollusks and native freshwater fishes (characids and catfish), and the introduction of a number of exotic marine species on an experimental basis. The production systems used are described in annex 6, Production Systems and Cultured Species in Latin America.

Commercial farming dominates export-led production of salmon, tilapia, oysters, and scallops destined for the United States, Canada, Europe, and Japan. In 2003, exports were valued at $4.6 billion, which were contributed to by salmon, shrimp, and tilapia (in that order). The economic impact of export-driven aquaculture is most evident in Chile, Ecuador, Mexico, and Colombia, where the industry has provided significant rural employment in production and processing. In common with the experience in Africa, however, the benefits of aquaculture have not accrued widely to the small-scale and subsistence fish farmers.

The main industry trends include (1) improving production efficiency to reduce costs and maintain market share; (2) expanding the export-driven processing industry; (3) diversifying to reduce risk and target new markets; (4) expanding marine aquaculture (including R&D investment in farming new marine species); (5) increasing production from freshwater aquaculture and of endemic species (characids) feeding low on the food chain; (6) expanding fish consumption in urban and rural areas; (7) allowing greater concern about sustainable practices; and (8) supporting regional cooperation.

Diagnosis

The diagnosis for Latin America shows similarities with that for Africa, but also shows some distinct differences:

- Aquaculture has undergone a significant recent development through the use of new technologies and commercial production systems. Commercial production is based largely on introduced species and technologies transferred through joint ventures or foreign expertise.

- Fish culture has not had a sustained impact on the livelihoods of impoverished farmers. As in Africa, subsistence aquaculture systems have had little sustainability, largely for the same reasons as in Africa, and because of low consumer demand for fish.
- Rapid growth followed the shift of development focus from the rural poor to middle-income and high-income (large-scale) farmers and to farming of species for export. SMEs benefited less from export-oriented growth, but their impact on domestic sales is increasing.
- Similar to the experience in Asia, commercial aquaculture succeeded first in countries with the institutional capacity to absorb and apply advanced technologies, and with supportive public policies (including for foreign investment) and proactive private sectors.
- The industry is a modest contributor to GDP, with economic impact most evident in Chile, Ecuador, Mexico, and Colombia. The main impact of the industry on rural livelihoods has been the direct and indirect generation of employment. The contribution to food fish supplies and local consumption is modest but rising. Supplies from capture fisheries have declined and there is a projected need for supply of an additional 2–3 million tons by 2020 (Delgado et al. 2003).
- The focus on foreign markets has accelerated the adoption of procedures and standards to guarantee quality and safety, and encouraged greater attention to environmental issues and production efficiency.
- Concentration on a few species and export markets makes the industry susceptible to price fluctuations, changes in currency exchange, economic situations in importing countries, competition and trade barriers, and developments in trade agreements. There is a need for diversification of species and markets to reduce risk.
- Prospects for growth are good, particularly in Brazil, because of the availability of natural resources (land and water), feed ingredients (fish meal and soybean), relatively cheap labor, supportive government policies, an interested private sector, and a thriving livestock industry.
- Main technical problems affecting the development of aquaculture are diseases, feed and seed availability, genetic deterioration of introduced species, and environmental issues.
- Aquaculture investment and expansion are constrained by complex laws and regulations and institutional bottlenecks. Enforcement is weak and development of information systems and capacity building is needed to support farmers and investors.

Regional Cooperation

The establishment of a regional aquaculture cooperative mechanism has received growing attention, and a feasibility study on the establishment of a single intergovernmental cooperation network is under examination.

The Case of Chile

Chile is an exception to much of the preceding description of aquaculture in Latin America. Chile is one of the largest world producers of salmon with exports worth $1.2 billion in 2003, accounting for about 6 percent of the country's exports. The seeding of a nascent sector by some foreign firms was reinforced by a strategic public-private partnership,[21] which facilitated the adaptation of superior foreign technologies and led to the development of a dynamic world-class export industry. Although half of the production capacity is now controlled by leading global firms, dependence on foreign materials and services has decreased with the growth of the industry and the evolution of local service industries.

As more firms entered the industry, the government's role changed from facilitator to regulator. Government organizations provide the regulatory framework for the issuance of permits, evaluation of EIAs, and surveillance of imported eggs based on a regulatory framework developed in the 1980s. The public sector continues to fund R&D, support local firms to keep abreast of global developments, and promote local knowledge to maintain a competitive edge. Annual research expenditure is about $10 million, a quarter of which comes from private firms (World Bank 2005).

The success of the salmon industry in Chile has been founded on several strategic advantages, including the following:

- A wide geographic range of farming locations and climatic conditions
- Pristine waters
- Low input and operational costs
- Stable economic policies
- Awareness, transfer, and adaptation of a proven superior technology
- Public-private partnerships
- Public policies and development services that attracted foreign investment from Norway, Japan, the United Kingdom, and elsewhere

CHAPTER SIX

Conclusions and Recommendations

> In comparison to other sectors of the world food economy . . . the fisheries and aquaculture sectors are poorly planned, inadequately funded, and neglected by all levels of government. [. . . yet . . .] fishing is the largest extractive use of wildlife in the world; and aquaculture is the most rapidly growing sector of the global agricultural economy.
>
> —*USAID SPARE Fisheries and Aquaculture Panel n.d.*

CONCLUSIONS

For more than two decades, aquaculture has grown at an average annual rate of 10 percent, confirming its claim as the global food production sector with the highest growth rate and filling a growing fish food supply gap that depleted wild fisheries are unable to bridge.

Importance

Aquaculture is important for numerous reasons, including the following:

- It accounts for 43 percent of the global fish food supply and is considered to be the primary source of any future increases in supply.
- More than 12 million people are directly employed in aquaculture, which reached 59.4 million tons with a farmgate value of $70.3 billion in 2004.

- Developing countries account for 90 percent of global aquaculture production.
- Aquaculture products are capturing an increasing share of the global fish trade, worth $71 billion in 2004, and more than 48 percent of the global fish trade is from developing countries.
- Aquaculture has a demonstrated capability to reduce poverty, improve livelihoods, and be a significant engine of growth.
- Unbridled aquaculture contributes to environmental degradation.
- Farming of aquatic species is inherently more efficient than livestock and has a smaller environmental footprint.
- Aquaculture provides a range of environmental services, has the potential to provide many more, and complements and integrates with rural development and coastal management.

Issues

Like any infant industry diversifying and expanding its horizons, the aquaculture sector competes with traditional users of land, water, and resources. It strains against traditional regulatory frameworks, designed without aquaculture in mind, and aggressively invades traditional markets for fish, provoking trade disputes and protectionism. In some respects, aquaculture is tracing the evolution of agriculture and livestock, intensifying and applying science and technology to increase productivity. Many of the issues confronting livestock and aquaculture are similar. These include the following:

- The framework for aquaculture governance is often deficient—policies, legislation, and strategic and physical plans often lag behind the investments—and expansion and adoption of BMPs is often sluggish.
- Pro-poor aquaculture policies and programs are emerging, but the pace of adoption remains slow.
- There are increasingly stringent environmental and sanitary standards, but enforcement of environmental regulations is often lax.
- There is a shortage of suitable sites and growing competition for water and scarce public goods, leading to conflicts among resource users in some culture systems.
- Access to technologies that enhance productivity (for example, disease control, feed industry, breeding programs) is often poor, largely as a result of human capacity and institutional deficiencies.
- Extension systems and services for smallholders are poorly supported and delivery mechanisms for pro-poor aquaculture are weak.
- Consumer demand is driving improved quality and food safety, while changing import standards and trade disputes limit the entry of aquaculture products into some of the more lucrative markets and distort markets and supplies.

- Civil society questions the impact of aquaculture on biodiversity and the environment.
- There is a body of disinformation on aquaculture, including on the nutritional quality of farmed fish.

Key Trends

Some key characteristics of modern aquaculture include:

- An aggressive market-driven expansion
- Increasing productivity based on intensification and structural change
- Emergence of a knowledge-based industry
- Competition for scarce water and land resources with other sectors
- Increasing environmental impacts (both positive and negative)

Chasing Changing Markets

The defining characteristics of the markets for aquaculture products include a shortened and more efficient production chain compared with capture fisheries; a convergence between domestic and export market standards; trade disputes; and industry consolidation and the weak market power of many small-scale producers. Timely delivery of standard products, vertical integration, and a more favorable cost structure makes aquaculture production increasingly more efficient than many capture fisheries. While the product chain is shortened through direct sales agreements between producers and giant retailers, the base is widened by development of specialized service providers, altering the patterns of value capture.

An Institutional Home

Because of its relative novelty as an economic sector, aquaculture lags behind other sectors in terms of policies, appropriate institutional and regulatory frameworks, and integration into development planning. While generally considered as a subsector of fisheries, aquaculture has far more in common with livestock and agriculture. Thus, institutionally, aquaculture often lies uncomfortably with capture fisheries, while fish, seaweed, and aquatic ecology lie equally uncomfortably with cattle, grains, and land management. The search for an institutional home for a complex and rapidly evolving sector that straddles a diversity of habitats, institutions, sciences, tenures, and production systems is among the many challenges facing development of sustainable aquaculture.

Science and Technology

Only in the last three decades have the concerted efforts of science and technology been directed toward improving productivity and management. Aquaculture lags well behind agriculture and livestock in this respect. This presents important opportunities, but demands an institutional framework to nurture the investments required to realize sustainable benefits. Rapid developments in seed production, fish feed technology and disease control, and the integration of aquaculture in both urban and rural economies are among the innovation domains driving aquaculture expansion. Recognition of the potential contribution of aquaculture to ecosystem services and the emergence of national and international norms, codes of practice, and standards for environmentally friendly aquaculture and healthy fish products are creating further opportunities for expansion.

Diversity and Productivity

An increasing number of aquatic species are being domesticated and cultured under increasingly diverse production systems. Although much attention has focused on intensive production of high-value species such as shrimp and salmon, the vast majority of aquaculture takes place at lower trophic levels with a small environmental footprint—such as plankton and plant feeders. There is a continuum from low-trophic level to high-trophic level production systems; for example, tilapia can be raised intensively and finds its way onto the menus of the finest restaurants, but it will yield modest returns from a backyard pond fertilized with garden compost.

Poor Infrastructure

The infrastructure needed to bring high-value, highly perishable products to markets, often from relatively isolated aquafarms, is frequently deficient. This infrastructure includes not only the transport "hardware," but also a suite of information and communications infrastructure providing traceability, market price information, and information on disease outbreaks and changing aquatic environmental conditions, such as red tides or impending floods.

Opportunities

Outside the nexus of Asian aquaculture, vast areas in coastal regions and river basins are suitable for aquaculture, particularly in SSA and Latin America. Past development efforts have, at best, yielded mixed results. A new chapter in aquaculture development will need to take full account of the lessons of past failures and apply the lessons learned in Asia and elsewhere in terms of business models and knowledge transfer. These lessons include building awareness of the need for: (1) effective public policies, planning, and governance; (2) a favorable investment climate; (3) involvement of stakeholders, enforcement of environmental controls, and transfer of technologies and skills; and (4) measures to mitigate negative externalities and unwanted social impacts.

International Demand

Developing countries have expressed the demand for aquaculture development in a range of forums, such as the FAO Committee on Fisheries, and in international and regional declarations—for example, the recent NEPAD *Fish for All* Summit Declaration and Plan of Action and the Resolution and Plan of Action of the Association of Southeast Asian Nations (ASEAN)/SEAFDEC Conference on Sustainable Fisheries for Food Security in the New Millennium.

The Pillars of Sustainable Aquaculture

Where aquaculture is already a robust and expanding sector, the challenges are those of equity, environmental sustainability, and trade. Where aquaculture is still an embryonic industry, the creation of an enabling environment is a further challenge. The road map for sustainable aquaculture is built on three main pillars, each of which is discussed in some detail in the following sections:

- Good governance, including establishment of an enabling environment for aquaculture investment through policies and practices, facilitating equitable access to water, land, resources, and markets
- A commitment to environmentally sustainable and healthy aquaculture
- Creation of the human and institutional capacity and knowledge required for management, innovation, and building of aquaculture infrastructure

Public and Private Sector Roles

The public sector has a vital role to play in creating an attractive investment climate, establishing a framework for disease control, monitoring transfer of live fish across boundaries, overseeing water management and environmental protection, ensuring the quality of feeds and seeds, and certifying the health and safety of aquaculture food products. Functions undertaken by the public sector may vary widely between countries, but include the following:

- The physical and environmental planning for coastal zones, wetlands, river basins, and water use
- Monitoring and enforcement of regulations and operation, or oversight of sanitary controls
- Allocation of leases over public waters in a transparent and equitable manner and arbitration on competing resource use
- Coordination of a participatory planning and policy review processes
- Interagency/ministry (and state/federal) coordination and policy coherence
- Support for core training and knowledge acquisition and innovation
- Provision of key infrastructure

Functions that may require public involvement include seed production and supply, maintenance of broodstock quality, and extension and certification schemes. Engagement of NGOs in the provision of microfinance, extension, and independent oversight of environmental and equity issues may be beneficial. Suitable incentives can enlist the private sector in the creation and operation of infrastructure, such as those for fish food safety and sanitary control, adapting proven technologies and pro-poor sustainable aquaculture models and encouraging partnerships in applied research among industry, government, and research institutions.

Environmental Sustainability

Introduction and continuance of environmentally friendly aquaculture systems is vital. Development assistance can foster adoption and application of codes and best practices for environmentally friendly aquaculture, which will increase economic returns while providing effective environmental stewardship and producing healthy products. A wide range of codes and best practices exist (see annex 2, Selected Codes, Instruments, and Tools for Responsible Aquaculture) setting out norms, standards, and guidelines in all major areas of concern, ranging from fish health to mangrove forest management. These knowledge items can be tailored to national needs, as undertaken recently in Vietnam with World Bank assistance. Among the actions required are the following:

- Provisions for undertaking the required environmental and social impact assessments
- Zoning of aquaculture and integration with coastal and river basin planning
- Measures to replace collection of wild seedstock
- Incentives to use processed feeds to reduce harvesting of trash fish for feeds and to use alternatives such as lysine-rich yeast and plant sources of essential nutrients
- Rigorous evaluation, risk assessment, and monitoring and control of species transfers and introductions between river basins, countries, and regions
- Provisions for monitoring of testing of water quality, and if necessary, preparation of the relevant water and product standards and development of human capacity in this field
- Use of ecolabels and certification systems, particularly to capture export markets
- Internalizing the environmental costs of aquaculture through fiscal and other measures

Knowledge and Human Capacity

Knowledge and human caacity are fundamental and investment in human and intangible capital is perhaps the highest priority for sustainable aquaculture.

This capital can be cost-effectively generated in developing countries through the use of networks and south-south cooperation backed by sustained support from the international community. The initiatives and approaches include joint ventures, formal and vocational and informal training, applied research alliances, and establishment of producer organizations. Investment in social capital through community-based management approaches and corporate links such as eChoupal in India provides access to financial capital, support infrastructure, and markets.

Public Interventions and Entry Points for Development Assistance

Many entry points for public interventions and development assistance have been illustrated in the preceding sections. The following examples are selected from this broad array.

Good Governance

In China and India, success was due to strategic planning and long-term efforts by government, including special institutional arrangements, supporting laws and regulations, and access to inputs, credit, and markets. Initiatives can include the following:

- Participatory preparation of a national aquaculture policy, development strategy and plan, and integration of the process with national economic plans, poverty reduction plans, environmental management plans, development assistance programs, and similar instruments
- Development of guidelines on process aspects of the multiagency and multistakeholder cooperation and cross-sector coordination in aquaculture
- Revision of aquaculture legislation, including land and water tenure rights, and policies to make it possible for the poor to engage in aquaculture production; for example, opening the way for leasing public land and water bodies to poor households
- Development of safeguards for community waters, wetlands, and similar common property
- Integration of aquaculture into coastal and river basin management plans

Economic Growth and Investment

Sector support can be considered along the entire supply and product value chain from underpinning knowledge industry and services to infrastructure and processing through the following:

- Preparing guidelines for private investment in aquaculture, including FDI

- Building incentives and processes for cooperation and partnerships among government agencies, private sector, producer groups, and NGOs for sustainable aquaculture development
- Mobilizing institutional credits for investment by poor people and women in aquaculture
- Designing incentives to stimulate aquaculture development in remote areas
- Facilitating reduction of risks through special provisions for infant industry

Corporate Aquaculture

Successful smallholder aquaculture is often built on the pioneering efforts of the larger entrepreneurs. The larger enterprises have the economies of scale to overcome logistics problems, can access finance and markets, and can secure the political patronage to reduce risk and cut through bureaucratic knots. By virtue of its access to knowledge on codes of responsible aquaculture, corporate aquaculture also has the obligation to implement industry norms and best practices. Similarly, banks and corporate sources of aquaculture finance have an obligation to ensure that corporate clients engage in responsible aquaculture practices, while the engagement of major food retailers in certification and ecolabeling schemes can reinforce and reward socially and environmentally responsible corporate aquaculture. Establishment of partnerships among major stakeholders at national and international levels would help create a level playing field for corporate producers and curtail incentives to "race to the bottom" in terms of environmental or social sustainability. Such partnerships could include IFIs, major wholesale or retailers, and custodians of codes of conduct for sustainable aquaculture, linking both the access to capital and revenues from production to the application of designated norms.

Poverty Reduction

A variety of pro-poor approaches have been illustrated in the preceding sections. A first step is to raise awareness among decision makers and international agencies with a view to include aquaculture in rural development and poverty reduction strategies. Additional specific pro-poor interventions can then be explored, such as:

- Creating diagnostics to design effective interventions
- Piloting proven approaches in other developing countries
- Integrating aquaculture into community development and rural development projects
- Fostering producer organizations

Environmental Management

The existing codes and guidelines can be blended into the development agenda, for example, by establishing specific safeguard policies for aquaculture investment to be used by client countries, development agencies, and commercial banks. Other interventions can include the following:

- Include aquaculture in improved land-use plans, water management, and irrigation and drainage projects to reduce risks like salinization, disease incidence, loss of biodiversity, and threats to wildlife populations.
- Where uncertainty and risk is high due to lack of information, support pilot projects to provide information to improve the information base, assess environmental effects, create stakeholder understanding, and support environmental protection.
- Create capacity and processes for effective monitoring and enforcement, including capacity to assess EIAs, and apply a precautionary approach based on risk assessment.
- Support processes for adoption of BMPs, development of certification, traceability, and ecolabeling schemes.
- Design environmental accounting systems for aquaculture and the accompanying fiscal measures to internalize social and environmental costs.

Trade

Aquaculture requires substantial public support to meet the stringent quality and safety standards of aquaculture products traded internationally. Establishment of veterinary controls, advisory services, and support for participation in trade fairs are among the public goods needed. International cooperation can help avert trade disputes and promote equitable trade.

Knowledge and Capacity Building

The lessons learned from technology transfer and capacity building in Asia described above can be applied in other regions. A key factor was the regional cooperation in training and knowledge sharing, providing a cost-effective solution through networks and processes without duplicating costly facilities and research in each country. The NACA model described above and detailed in annex 4 can be adapted and molded to meet the requirements of other regions.

Infrastructure

Providing for aquaculture in the design of rural road networks, flood control systems, and irrigation and drainage infrastructure will contribute to an effective enabling environment, reducing costs of water and resource management, creating the environmental space for aquaculture development, and engaging aquaculture as a contributor to environmental services. The infrastructure requirement includes not only roads, water control systems, and energy, but also a suite of information and communications infrastructure providing traceability, market price information, or information on disease outbreaks and changing aquatic environmental conditions, such as red tides, or impending floods.

RECOMMENDATIONS

Following is a summary of recommendations made for actions by client countries and the development community:

1. Raise **awareness** of the potential of aquaculture for economic growth, poverty alleviation, food security, and environmental services. Concurrently stress the need for good governance to counter the threats posed to the environment and the poor by unsustainable aquaculture practices.
2. Improve aquaculture **governance** through informed policies based on science-based diagnostics and coordinated public and private activities and working from a plan formulated through participation and consultation.
3. Create an **enabling environment** for private sector investment, innovation, and expansion at both corporate and smallholder levels.
4. Adopt and apply recognized codes of responsible aquaculture and best practices to ensure **environmental sustainability** of aquaculture, take measures to internalize the environmental costs, and use market mechanisms such as certification and ecolabeling to forge an economic backbone for and social awareness of environmentally friendly and healthy aquaculture.
5. Provide **sustained public support for pro-poor aquaculture** using proven models adapted to local realities with a focus on commercially viable culture systems, or integration into rural livelihoods and agricultural practices.
6. Create national and regional **knowledge networks** to transfer and adapt proven technologies, to cost-effectively build human and institutional capacity and support south-south technology transfer and investment.
7. Create aquaculture **infrastructure,** not only in terms of roads and other "hardware," but also in terms of communications, institutions, and support services.
8. Assess the increasing threat to **biodiversity,** including loss of wild germplasm, posed by aquaculture and consider means of countering these threats, including the rigorous application of core biodiversity-related guidelines.

9. Through collaboration between client countries and development partners, prepare specific **safeguard policies** and guidelines for aquaculture for consideration by international financial institutions with a view to extending such norms through the Equator Principles.
10. Establish institutional and informal **partnerships** to support agreed agendas for sustainable aquaculture in developing countries, including initiatives to avert trade disputes and address biodiversity loss.

ANNEXES

Definitions of Aquaculture Production Systems

Barrages: Semipermanent or seasonal enclosures formed by impervious man-made barriers and appropriate natural features.

Cages: Open or covered enclosed structures constructed with net, mesh, or any porous material allowing natural water interchange. These structures may be floating, suspended, or fixed to the substrate but still permitting water interchange from below.

Capture-based aquaculture: The practice of collecting "seed" material—from early life stages to adults—from the wild, and its subsequent on-growing in captivity to marketable size, using aquaculture techniques.

Enclosures and pens: Water areas confined by net, mesh, and other barriers allowing uncontrolled water interchange and distinguished by the fact that enclosures occupy the full water column between substrate and surface; pens and enclosures will generally enclose a relatively large volume of water.

Enhancement: Any activity aimed at supplementing or sustaining the recruitment of, or improving the survival and growth of, one or more aquatic organisms, or at raising the total production or the production of selected elements of the fishery beyond a level that is sustainable by natural processes. It may involve stocking, habitat modification, elimination of unwanted species, fertilization, or combinations of any of these practices.

Extensive: Production system characterized by (1) a low degree of control (for example, of environment, nutrition, predators, competitors, disease agents);

(2) low initial costs, low-level technology, and low production efficiency (yielding no more than 500 kg/ha/yr); (3) high dependence on local climate and water quality; and (4) use of natural water bodies (for example, lagoons, bays, embayments) and of natural, often unspecified, food organisms.

Hatcheries: Installations for housing facilities for breeding, nursing, and rearing seed of fish, invertebrates, or aquatic plants to fry, fingerlings, or juvenile stages.

Hyperintensive: System of culture characterized by a production averaging more than 200 tons/ha/yr, by the use of a complete (processed) fully formulated feed to meet all diet requirements of the species, stocking with hatchery-reared fry, no fertilizers used, full predator and antitheft precautions taken, highly co-coordinated and controlled regimes, usually pumped or gravity-supplied water or cage-based, full use of water exchange and aeration with increasing levels of control over supply and quality, usually in flowing water ponds, cage systems, or tanks and raceways.

Integrated agriculture-aquaculture (IAA): Semi-intensive aquaculture systems in synergy with agriculture (including animal husbandry), in which an output from one subsystem, which otherwise may be wasted, is used as an input to another subsystem, resulting in a greater efficiency of output of desired products from the land/water farm area. Other forms of integration using the same principle of waste recycling include intensive marine coculture of fish/shrimp-shellfish-seaweeds, aquaculture in heated effluents, sewage-fed aquaculture, and integrated mangrove-shrimp farming.

Intensive: System of culture characterized by (1) a production of up to 200 tons/ha/yr; (2) a high degree of control; (3) high initial costs, high-level technology, and high production efficiency; (4) tendency toward increased independence of local climate and water quality; (5) use of manmade culture systems.

Nurseries: Refer generally to the second phase in the rearing process of aquatic organisms and to small, mainly outdoor ponds and tanks.

Polyculture: The rearing of two or more noncompetitive species in the same culture unit.

Ponds and tanks: Artificial units of varying sizes constructed above or below ground level capable of holding and interchanging water. Rate of exchange of water is usually low, that is, not exceeding 10 changes per day.

Raceways and silos: Artificial units constructed above or below ground level capable of high rates of water interchange in excess of 20 changes per day.

Rafts, ropes, and stakes: The culture of shellfish, notably mussels, and seaweeds usually conducted in open waters using rafts, long lines, or stakes. The stakes are impaled in the seabed in intertidal areas and ropes are suspended in deeper waters from rafts or buoys.

Rice-cum-fish paddies: Paddy fields used for the culture of rice and aquatic organisms; rearing them in rice paddies to any marketable size.

Sea-ranching: The harvest of enhanced capture fisheries, that is, the raising of aquatic animals, mainly for human consumption, under extensive production systems, in open space (oceans, lakes) where they grow using natural food supplies. These animals may be released by national authorities and recaptured by fishermen as wild animals, either when they return to the release site (salmon), or elsewhere (seabreams, flatfishes).

Semi-intensive: Systems of culture characterized by a production of 2 to 20 tons/ha/yr, which are dependent largely on natural food, which is augmented by fertilization or complemented by the use of supplementary feed, stocking with hatchery-reared fry, regular use of fertilizers, some water exchange or aeration, and often pumped or gravity-supplied water, and normally in improved ponds, some enclosures, or simple cage systems.

Semi-extensive: System of culture characterized by a production of 0.5–5 tons/ha/yr, possibly supplementary feeding with low-grade feeds, stocking with wild-caught or hatchery-reared fry, regular use of organic or inorganic fertilizers, rain or tidal water supply, and/or some water exchange, and simple monitoring of water quality, and normally in traditional or improved ponds; also some cage systems, for example, with zooplankton feeding for fry.

Source: FAO 1990; FAO Aquaculture Glossary http://www.fao.org/fi/glossary/aquaculture/.

ANNEX TWO

Selected Codes, Instruments, and Tools for Responsible Aquaculture

INTERNATIONAL AND REGIONAL CODES

FAO Code and Technical Guidelines

FAO Code of Conduct for Responsible Fisheries. Available at http://www.fao .org/documents/show_cdr.asp?url_file=/DOCREP/005/v9878e/v9878e00.htm

FAO Technical Guidelines for Responsible Fisheries:

- Aquaculture Development, 1997. Available at ftp://ftp.fao.org/docrep/fao/ 003/W4493e/W4493e00.pdf
- Aquaculture Development 1. Good Aquaculture Feed Practice, 2001. Available at ftp://ftp.fao.org/docrep/fao/005/y1453e/y1453e00.pdf
- Integration of Fisheries into Coastal Area Management, 1996. Available at ftp://ftp.fao.org/docrep/fao/003/W3593e/W3593e00.pdf
- Precautionary Approach to Capture Fisheries and Species Introductions, 1996. Available at http://www.fao.org/documents/show_cdr.asp?url_file=/ DOCREP/003/W3592E/W3592E00.htm

Responsible Fish Utilization, 1998. Available at ftp://ftp.fao.org/docrep/fao/ 003/w9634e/w9634e00.pdf

Guidelines on the Collection of Structural Aquaculture Statistics, 1997

Species Introductions and Biodiversity

Cartagena Protocol on Biosafety. Available at http://www.biodiv.org/biosafety/ protocol.asp

Convention on Biological Diversity (CBD), 1992. Available at http://www.biodiv
.org /convention/articles.asp

Convention on International Trade in Endangered Species of Wild Fauna and
Flora (CITES), 1973. Available at http://www.cites.org/

EIFAC Code of Practice and Manual of Processes for Consideration of Introductions and Transfers of Marine and Freshwater Organisms, 1988, European Inland Fisheries Advisory. Available at http://cdserver2.ru.ac.za/
cd/011120_1/Aqua/ SSA/codes.htm

FAO Technical Paper, International Introductions of Inland Aquatic Species,
1988. Available at http://www.fao.org/docrep/x5628E/x5628e00.htm#
Contents

International Council for the Exploration of the Seas (ICES)/ European Inland
Fisheries Advisory Commission (EIFAC) Code of Practice on the Introductions and Transfers of Marine Organisms, 2004. Available at http://www
.ices.dk/reports/general/2004/ICESCOP2004.pdf

Health Management and Best Practice

Better-Practice Approaches for Culture-Based Fisheries Development in Asia,
2006. Available at http://www.aciar.gov.au/web.nsf/att/ACIA-6M98FT/$file/
CBF_manual.pdf

CODEX Alimentarius. Available at http://www.codexalimentarius.net/web/
standard_list.do?lang=en

Development of HARP Guidelines. Harmonised Quantification and Reporting
Procedures for Nutrients. SFT Report 1759/2000. TA-1759/2000. ISBN
82-7655-401-6. Available at http://www.sft.no/publikasjoner/vann/1759/
ta1759.pdf

FAO/NACA Asia Diagnostic Guide to Aquatic Animal Diseases, 2001. Available
at http://www.fao.org/documents/show_cdr.asp?url_file=/DOCREP/005/
Y1679E/Y1679E00.HTM

FAO/NACA Asia Regional Technical Guidelines on Health Management for
the Responsible Movement of Live Aquatic Animals and the Beijing Consensus and Implementation Strategy, 2000. Available at http://www
.fao.org/documents/ show_cdr.asp?url_file=/DOCREP/005/X8485E/x8485
e02.htm

FAO/NACA Manual of Procedures for the Implementation of the Asia Regional
Technical Guidelines on Health Management for the Responsible Movement
of Live Aquatic Animals. 2001. Available at http://www.fao.org/documents/
show_ cdr.asp?url_file=/DOCREP/005/Y1238E/Y1238E00.HTM

Holmenkollen Guidelines for Sustainable Aquaculture, 1998. Available at http://
www.ntva.no/rapport/aqua.htm

International Aquatic Animal Health Code, 2005. Available at http://www.oie
.int/ eng/normes/fcode/a_summry.htm

Shrimp Culture

Bangkok FAO Technical Consultation on Policies for Sustainable Shrimp Culture, Bangkok, Thailand, December 8–11, 1997. Available at http://www.fao.org/documents/show_cdr.asp?url_file=/DOCREP/006/x0570t/x0570t00.HTM

Codes of Practice and Conduct for Marine Shrimp Aquaculture, 2002. Available at http://www.fw.vt.edu/fisheries/Aquaculture_Center/ Power_Point_Presentations/FIW%204514/Lecture%209.1%20-%20aquaculture%20and%20environment/shrimpCOP.pdf

Codes of Practice for Responsible Shrimp Farming. Available at http://www.gaalliance.org/code.html#CODES

Code of Practice for Sustainable Use of Mangrove Ecosystems for Aquaculture in Southeast Asia, 2005. Available at http://www.ices.dk/reports/ general/2004/ICESCOP2004.pdf

The International Principles for Responsible Shrimp Farming, in preparation. Available at http://www.enaca.org/modules/mydownloads/singlefile.php?cid=19&lid=755

Report of the Ad-hoc Expert Meeting on Indicators and Criteria of Sustainable Shrimp Culture, Rome, Italy. April 28–30, 1998. Available at http://www.fao.org/ documents/show_cdr.asp?url_file=/DOCREP/006/x0570t/x0570t00.HTM

NATIONAL CODES AND BEST PRACTICES

Canada: National Code on Introductions and Transfers of Aquatic Animals, 2003. Department of Fisheries and Oceans, Government of Canada. Available at http://www.dfo-mpo.gc.ca/science/aquaculture/code/Code2003_e.pdf

Chile: Code of Good Environmental Practices (CGEP) for Well-Managed Salmonoids Farms, 2003. Available at http://library.enaca.org/certification/publications /Code_2003_ENGLISH.pdf

India: Guidelines for Sustainable Development and Management of Brackish Water Aquaculture, 1995. Available at http://www.mpeda.com/

Japan: Basic Guidelines to Ensure Sustainable Aquaculture Production, 1999.

Philippines: Fisheries Code, 1998. Available at http://www.da.gov.ph/ FishCode/ra8550a.html

Scotland: Code of Practice to Avoid and Minimise the Impact of Infectious Salmon Anaemia (ISA), 2002. Available at http://www.marlab.ac.uk/FRS.Web/Uploads/ Documents/ISACodeofPractice.pdf

Sri Lanka: Best Aquaculture Practices (BAP) for Shrimp Framing Industry in Sri Lanka. Available at http://www.naqda.gov.lk/pages/BestAquaculture Practice Methods.htm

Thailand: Thailand Code of Conduct for Shrimp Farming (in Thai). Available at http://www.thaiqualityshrimp.com/coc/home.asp

United States: Code of Conduct for Responsible Aquaculture Development in the U.S. Exclusive Economic Zone, 2002. Available at http://www.nmfs .noaa.gov /trade/AQ/AQCode.pdf

Guidance Relative to Development of Responsible Aquaculture Activities in Atlantic Coast States, 2002. Available at http://www.asmfc.org/publications/ special Reports/aquacultureGuidanceDocument.pdf

Guidelines for Ecological Risk Assessment of Marine Fish Aquaculture, 2005. NOAA Technical Memorandum NMFS-NWFSC-71. Available at http:// www.nwfsc.noaa.gov/assets/25/6450_01302006_155445_NashFAOFinal TM71.pdf

U.S. Department of Agriculture Aquaculture Best Management Practices Index, 2004. Available at http://efotg.nrcs.usda.gov/references/public/AL/ INDEX.pdf

INDUSTRY/ORGANIZATION CODES

Australian Aquaculture Code of Conduct Available at http://www.pir .sa.gov.au/ byteserve/aquaculture/farm_practice/code_of_conduct.pdf

British Columbia Salmon Farmers Association (BCSFA) Code of Practice, 2005. Available at http://www.salmonfarmers.org/pdfs/codeofpractice1.pdf

A Code of Conduct for European Aquaculture. Available at http://www.feap .info/FileLibrary/6/CodeFinalD.PDF

Draft Protocol for Sustainable Shrimp Production, in preparation. Available at http://www.ntva.no/rapport/aqua.htm

Environmental Code of Practice for Australian Prawn Farmers, 2001. Available at http://www.apfa.com.au/prawnfarmers.cfm?inc=environment

Judicious Antimicrobial Use in U.S. Aquaculture: Principles and Practices, 2003. Available at http://www.nationalaquaculture.org/pdf/Judicious% 20Antimicrobial%20Use.pdf

New Zealand Mussel Industry Environmental Codes of Practice, 2002. Mussel Industry Council Ltd., Blenheim.

ENVIRONMENT AND SOCIAL SAFEGUARD POLICIES AND GUIDELINES

World Bank Group

Environmental Assessment Sourcebook, 1991. Available at http://web.worldbank .org/WBSITE/EXTERNAL/TOPICS/ENVIRONMENT/EXTENVASS/0,, contentMDK:20282864~pagePK:148956~piPK:216618~theSitePK: 407988,00.html

The Environmental and Social Review Procedure (ESRP), 2006. Gives direction to IFC officers in implementing the Policy on Social and Environmental Sustainability and reviewing compliance and implementation by private sector projects. Available at http://www.ifc.org/ifcext/enviro.nsf/ Attachments ByTitle/pol_ESRP2006/$FILE/ESRP2006.pdf

Fish Processing Guideline, 1998. Available at http://www.ifc.org/ifcext/enviro .nsf/AttachmentsByTitle/gui_fishproc/$FILE/fishprocessing.pdf

General Environmental Guideline, 1998. Available at http://www.ifc.org/ ifcext/ enviro.nsf/AttachmentsByTitle/gui_genenv_WB/$FILE/genenv_PPAH.pdf

The Policy on Social and Environmental Sustainability, 2006. Defines IFC's responsibility for supporting project performance in partnership with clients. Available at http://www.ifc.org/ifcext/enviro.nsf/Content /SustainabilityPolicy

Pollution Prevention and Abatement Handbook (PPAH), 1998. Available at http://www.ifc.org/ifcext/enviro.nsf/Content/PPAH

World Bank's 10 Environment and Social Safeguard Policies, designed to assist Bank staff as they apply the safeguard policies and procedures. Available at http://web.worldbank.org/WBSITE/EXTERNAL/PROJECTS/EXTPOLICIES/ EXTSAFEPOL/0,,menuPK:584441~pagePK:64168427~piPK:64168435~the SitePK:584435,00.html

Inter-American Development Bank

Environment and Social Safeguard Compliance Policies, 2006. Available at http:// www.iadb.org/IDBDocs.cfm?docnum=665902

Procedure for Environmental and Labor Review of IIC Projects by the Inter-American Investment Corporation (IIC), 1999. Available at http://www.iic .int/Policies/gn1293.ASP

Asian Development Bank

Environment Assessment Guidelines, 2003. Available at http://www.adb.org/ Documents/Guidelines/Environmental_Assessment/default.asp

Environment Policy, 2002. Available at http://www.adb.org/Environment/ default.asp

Indigenous Peoples (IP) Policy, 1998. Available at http://www.adb.org/Indigenous Peoples/default.asp

Involuntary Resettlement (IR) Policy, 1995. Available at http://www.adb.org/ Resettlement/default.asp

Import Risk Analysis

Asia-Pacific Economic Cooperation: Capacity and Awareness Building on Import Risk Analysis (IRA) for Aquatic Animals. Proceedings of the workshops held April 1–6, 2002, in Bangkok, Thailand, and August 12–17, 2002, in Mazatlan, Mexico. APEC FWG 01/2002, NACA, Bangkok. Available at http://www.enaca.org/modules/mydownloads/visit.php?cid=21&lid=528& PHPSESSID=f0138bb49acd5570e224a09e9e808cea

Manual on Risk Analysis for the Safe Movement of Aquatic Animals, 2002. Available at http://www.apec.org/apec/publications/all_publications/fisheries _working.MedialibDownload.v1.html?url=/etc/medialib/apec_media_ library/downloads/workinggroups/fwg/pubs/2004.Par.0001.File.v1.1

Australia: Generic Import Risk Analysis (IRA) of Non-viable Non-salmonid Freshwater Finfish, 2002. Available at http://www.daff.gov.au/corporate_ docs/publications /pdf/market_access/biosecurity/animal/2002/2002-19a.pdf

Import Risk Analysis Handbook, 2003. Available at http://www.affa.gov.au/ corporate_docs/publications/pdf/market_access/biosecurity/bde/ira_ handbook_revised.pdf

ANNEX THREE
Portfolio Analysis

Table A3.1 Portfolio of World Bank Projects with an Aquaculture Component (as of May 2006)

Country	Project Name	Cost[a]	Aquaculture Components [Outcome of Project Completion Review]
Albania	Pilot Fishery Development Project (2002–07)	2.1	One of the main objectives of the project is to restore the country's previous capacity in aquaculture and explore the potential for further development of aquaculture, particularly for high-value species. [Ongoing]
Bangladesh	Oxbow Lakes Fisheries Project (1979–86)	7.7	Introduction of new fish seed production.
Bangladesh	Shrimp Culture Project (1986–93)	27.4	Intensifying the production of brackish-water shrimp. [Satisfactory]
Bangladesh	Forth Fisheries (1999–)	14.2	The components include development and management of coastal shrimp aquaculture and freshwater aquaculture extension and training. [Ongoing]
China	Rural Credit Project II (1985–91)	96.8[b]	No information
China	Freshwater Fisheries Project (1986–92)	122.5	Freshwater fish culture in eight major urban centers were developed; area under intensive fish cultivation has been expanded. [Satisfactory]
China	Coastal Lands Development Program (1988–94)	138.8	Development of coastal culture fisheries (eels, shrimp, laver) achieved but encountered disease problems. [Satisfactory]
China	Rural Credit Project III (1988–94)	28	Establish and rehabilitate to grow fry and commercial fish using surface water of lakes and rivers. [Satisfactory]
China	Rural Credit Project IV (1990–95)	65.6	Shrimp farm rehab, scallops, seaweed, freshwater fish with support facilities. [Unsatisfactory]
China	Jiangxi Agriculture Development Project (1991–96)	15.1	Development and intensification of freshwater fish culture, including fish, crabs, and freshwater pearls. [Satisfactory]
China	Shaanxi Agricultural Development Project (1989–97)	22.7	Establishment of new fishponds. [Satisfactory]

Country	Project name		Description
China	Hebei Agricultural Development Project (1990–97)	41.2	Rehabilitation of shrimp ponds and constriction of new fishponds, scallop production, supports for hatcheries, feed mills, and freezers/cold stores. [Satisfactory]
China	Guangdong Agricultural Development Project (1991–98)	43.3	Construction of brackish-water fishponds, freshwater fishponds, shrimp and oyster cultivation, and support facilities including feed mills and hatcheries. [Satisfactory]
China	Henan Agricultural Development Project (1991–99)	18.1	Construction of 1,426 ha freshwater fishponds and support facilities including feed mills. About 3,600 farmers and technicians have been trained. [Satisfactory]
China	Songliao Plain Development Project (1998–2003)	26.8	Shrimp culture rehabilitation; establishment of scallop culture, clam, and river crabs hatcheries. [Satisfactory]
China	Southwest Poverty Reduction Project (1995–2004)	37.5[b]	Freshwater shrimp, marine shrimp, marine pearl, clams, cage fish culture. [Outcome not available]
China	Shanxi Poverty Reduction Project (1996–2004)	16.7	Fishpond improvement and aquaculture supporting activities. [Outcome not available]
China	Heilongjiang Agricultural Development (1997–2004)	4.8	Construction of fishponds. [Satisfactory]
China	Sustainable Coastal Resources Development Project (1998–2006)	146	Coastal zone management, marine aquaculture (clams, oysters, scallops, yellow croaker, other species), shrimp farm rehabilitation, and seafood processing. [Ongoing]
Egypt, Arab Rep. of	Fish Farming Development Project (1981–89)	33.3	Construction of fishponds with technical assistance and training. Five years' delay in implementation. [Partially satisfactory]
Ghana	Fisheries Sub-Sector Capacity Building Project (1995–2005)	1.5	Promotion of private sector investment in aquaculture, establishment of fingerling production centers (tilapia, catfish), development of a pilot aquaculture center. [Satisfactory]
India	Inland Fisheries Project (1979–88)	66.9	The project increased production of carp using underused ponds and then developed a carp seed industry in India.

(continued)

Table A3.1 (Continued)

Country	Project Name	Cost[a]	Aquaculture Components [Outcome of Project Completion Review]
India	Shrimp and Fish Culture Project (1992–2001)	30.6	Developed brackish-water shrimp culture (eighty-five percent of project costs).
Indonesia	Fisheries Support Services Project (1986–94)	4.2	Rehabilitation of fishponds (*tambaks*): improve supply of shrimp seed. [Unsatisfactory]
Kenya	Fisheries Project (1980–87)	2.1[b]	The proposed Fish Farming Development Center was never constructed; fish farmers were never trained. [Unsatisfactory]
Malawi	Fisheries Development Project (1991–2000)	0.0	Developing pilot fish-farming models to help integrate aquaculture into crop-farming systems and rehabilitating and developing existing capture fisheries were proposed. The aquaculture subcomponent was cancelled. [Unsatisfactory]
Mexico	Aquaculture Development Project (1994; terminated)	58[b]	The project focused on social sector aquaculture producers in seven states and some public sector interventions. The project was terminated at the government's request after nearly two years of operation from the date of loan effectiveness.
Philippines	Fisheries Credit Projects (1974–79)	5.2	Rehabilitation and improvement of brackish-water ponds. [Satisfactory]
Vietnam	Hon Mun Marine Protected Area Pilot Project (2001–05)		Small-scale seaweed culture with some other trials. [Ongoing]
Vietnam	Coastal Wetlands Protection and Development Project (2000–06)	15.9	Coastal pond aquaculture (shrimp, crabs, and others) and clam culture. [Ongoing]

Sources: World Bank ICRs and other Bank sources.

a. When the actual cost of aquaculture component is not available, either estimated cost at the appraisal or the cost of larger component was indicated.

b. Values are estimated costs at appraisal.

Table A3.2 IFC Aquaculture Projects 1992–2006 (in US$ million)

Country	Project Name	Total Cost	IFC Loan	Description of the Project
Madagascar	Aquaculture de Crevettes de Besalampy (Approved May 2001)	72.0	16.0	Following the success of Aqualma, the project expanded the sponsors' shrimp-farming activities, which consist of an integrated semi-intensive 4,400 tons/annum shrimp production and processing operation.
Madagascar	Aquaculture de Mahajamba (Aqualma) 1992–96		6.4	This is a pioneer shrimp-farming project established in 1992. By 1998, Aqualma was the only shrimp farm in the country with a significant contribution to the shrimp sector and the economy at large: it accounted for 17 percent of the country's total shrimp production, 19 percent of the total shrimp export volume, and 11 percent of the employment in the shrimp sector.
Madagascar	Les Gambas De L'ankarana (Approved January 2004)	27.9	6.5	The project established an integrated semi-intensive shrimp farm.
Venezuela, R. B. de	Inter Sea Farms de Venezuela, C.A. Phase I & II (Approved June 2000)		8.0	Phase I involves the expansion of the existing hatchery, the construction of about 1,000 ha of ponds, and all farm infrastructure and equipment. Phase II consists of the processing facilities and an additional 1,500 ha of ponds. [Ongoing]
Honduras	Grupo Granjas Marinas S.A. de C.V. (Approved April 1998)	18.7	6.0	Expansion of its shrimp postlarva hatchery, farming, and processing sites. The development outcome was rated as "Excellent" for business success and economic sustainability, and "Satisfactory" for environmental effects and private sector development. FRR =17 percent, ERR=20 percent

(continued)

Table A3.2 (Continued)

Country	Project Name	Total Cost	IFC Loan	Description of the Project
Belize	Nova Companies (Belize) Ltd. and Ambergris Aquaculture Ltd. (Approved May 1998)	15.2	6.0	Expansion of an existing shrimp production operation, and construction of a hatchery. Development outcome was rated as "Satisfactory" for the most categories, but "Partly unsatisfactory" for environmental effects. FRR=11.6 percent, ERR=11.6 percent
Indonesia	Central Pertiwi Bahari (Pipeline)		45.0	In January 2006, IFC announced an agreement to provide a loan facility of up to $45 million to PT. Central Pertiwi Bahari, a subsidiary of Charoen Pokhpand Group. CPB is Indonesia's leading integrated shrimp operator and a major exporter of shrimp products.
China	Nantong Wangfu Special Aquatic Products Co., Ltd (Pipeline)		19.0	In January 2006, IFC signed its first investment agreement in China's agribusiness sector to provide approximately $19 million in financing to Nantong Wangfu Special Aquatic Products Co., Ltd. The financing will help to implement an eel farming and processing project in Nantong, in the Jiangsu province of China.

Source: IFC project database.

Note: ERR = economic rate of return; IFC = International Finance Corporation; FRR = financial rate of return; €1 = US$1.2 (approx.).

Development agency aquaculture portfolios: In addition to the examination of the World Bank Group portfolio, lessons and experiences were drawn from evaluations, completions reports, and ongoing projects by the African Development Bank, European Union, Food and Agriculture Organization, United Nations Development Programme, World Wildlife Fund, UD Department for International Development, Danish International Development Agency, Japan International Cooperation Agency, German Agency for Technical Cooperation, Norwegian Agency for Development Cooperation, Canadian International Development Agency, Belgium Aid (DGCI), Swedish International Development Cooperation Authority, Finnish International Development Agency (FINNIDA), U.S. Peace Corps, Centre d'excellence-production, innovation et développement / Institut de recherche agricole pour le développement, CCTA (Netherlands), International Food Policy Research Institute, International Fund for Agriculture Development, U.S. Agency for International Development, International Center for Living Aquatic Resource Management Global Environmental Facility, Network of Aquaculture Centers in Asia-Pacific, Collaborative Research Support Program, Australian Centre for International Agricultural Research, Australian Government Overseas Aid Program, Association for the Development of Fish-farming and Fishing in Rhône-Alpes, European Commission, and Norwegian College of Fishery Science. Two studies were of particular note: ADB 2004 and Braga and Zweig 1998. The portfolio review included examination of case studies and documents on aquaculture and coastal management in African countries that include documents about and from the Arab Republic of Egypt, Nigeria, Madagascar, Mozambique, Tanzania, Ghana, Zambia, South Africa, the Democratic Republic of Congo, Cameroon, Côte d'Ivoire, The Gambia, Uganda, Zimbabwe, Tunisia, and Brazil.

Wealth Creation and Poverty Alleviation— The Asian Experiences

T his annex reviews selected pro-poor aquaculture and community aquaculture-based fisheries initiatives, projects, programs and supporting policies in several Asian countries.

AQUACULTURE FOR RURAL LIVELIHOODS

Drivers of change—markets for high-value fish. Export-led demand has changed the structure of Asia's aquaculture. Trade has focused on three major markets: the United States, Japan, and the European Union; but it is projected to shift to increasing south-south trade (Delgado et al. 2003), particularly among Asian countries. Species and culture systems have changed to meet changing export and domestic consumer preferences. Not only have the traditionally fish-eating Asian countries, such as China, Bangladesh, Vietnam, and Thailand, increased and diversified aquaculture production, but countries with a relatively low level of fish consumption, such as India, have joined the ranks of major producers and exporters.

Trade disputes. Trade remains a volatile area of tension between developed and developing countries, and between the rich and the poor. The complexities of food safety and public health concerns and other technical barriers have had dramatic effects on market access for Asian countries. The impacts have had a disproportionate effect on small producers and smaller economies because of economies of scale in the cost structure of HACCP and SPS compliance regimes (Ahmed 2005). Conversely, the elimination of harmful tariffs (tariff escalation and tariff peaks), such as tariffs that escalate with the level of processing, can result in important gains for poor people involved in input supply value added activities (Ahmed 2004).

Technology. Technology has transformed Asian aquaculture from a subsistence food production system to a major agribusiness industry. Hatchery technology, pelleted feeds, and disease control have fostered intensification. On-farm fish production is only one link in the aquaculture value chain, accounting for perhaps less than one-half of the total value added by industry. Modern fish-farming technology has expanded beyond the traditional ponds to rice fields, floodplains, rivers, and coastal waters, supplying vast quantities of food fish to growing domestic and international markets.

There has been a growing coordination of private sector input and output chains, including formal and informal links between smallholder producers and large processing companies, leading the industry toward greater efficiency, better quality assurance, secure margins for producers, and competitive prices for products. Export certification schemes have further streamlined production, processing, distribution, and retail chains. For one species after another (shrimp, catfish, tilapia), the product chains are increasingly molded by urban consumption behavior, as supermarket chains force the timely supply of quality products in domestic and export markets. The growing appetite for new cultured species (grouper, seabass, and Nile perch) means that it is probably only a matter of time and technology before these species follow the route of salmon in terms of productivity, price, and production.

Economic and social transformation. The combined effects of market demand, technological innovations, and infusions of corporate capital contributed to changes in the scale and business models in aquaculture operations. The new trends have given rise to more intensive production practices, forcing changes in the industrial organization of aquaculture. As individual farms linked to more organized input and output markets, consolidation occurred along the entire supply chain. In the Mekong delta, for example, this consolidation delivered economies of scale, greater efficiency, and major increases in catfish production as well as in the number of households involved in aquaculture (Monti and Crumlish n.d.).

China

Since the liberalization of the economy in the mid-1980s, aquaculture development has played a significant role in increasing the income of millions of poor farmers, and has worked as an engine for rural economic growth. In poor and remote provinces of China, initial adoption and benefits of aquaculture came through investment in integrated farming of fish in rice paddies to produce staple fish, such as Chinese carp, followed by gradual intensification and changes in the combination of species cultured to supply diversifying markets.

Between 1980 and 2000, aquaculture created 3 million full-time jobs in rural areas out of a total of 5.6 million jobs in fisheries (capture and culture) (see figure A4.1). Its contribution to full-time employment in the fisheries sector rose from 45 percent to 67 percent in the same period (Li 2003). The average per capita income of the labor force in fisheries and aquaculture also rose from Yuan (Y) 171 to 4,323 during 1980–98 (Wang 2001). In addition, the expansion of aquaculture has enabled the development of aquaculture-related industries in rural areas, which have provided significant additional employment opportunities (see box A4.1).

Rice-fish culture helping millions of farmers in rural China. Rice-fish culture is a relatively easy, low-cost, low-risk entry point for rural farming communities to improve their livelihoods without jeopardizing the sustainability of rice production. China's age-old tradition of growing fish on paddy fields was revived in the 1980s following economic reforms that liberalized land use and farm management systems. The practice emerged as an important contributor to poverty reduction and rural economic growth, especially in remote, poorer provinces and regions. It evolved from traditional, extensive family operations to medium-to-large or commercial operations in some provinces. The rapid expansion was the result of growing interest among farmers, coupled with sustained and comprehensive public support, including extension and advisory services, policy incentives, and better accessibility to loans for renovation of conventional paddy field to suit fish culture. Rice-fish culture has been incorporated into the overall rural and agriculture development plans by many local governments.

Figure A4.1   Growth in Fisheries Employment in China, 1974–2000

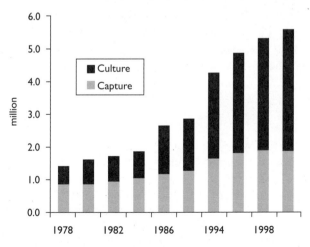

Source: Li 2003.

The Freshwater Fisheries Project developed integrated fish-farming complexes around eight major Chinese cities, improving existing ponds, using land unsuitable for crops for new pond construction, and introducing technologies associated with more advanced fish culture, such as artificial feeds and water quality controls. The project provided a transition stage between the traditional carp polyculture and modern fish farming, and paved the way for the introduction of intensive monoculture systems. The project has served as a model for the establishment of commercial fish farms and is regarded as highly successful by the borrower and the World Bank. The main reasons for this are as follows:

- Simple project design benefiting from a two-year preparation phase leading to sound technical and institutional design
- Peri-urban location providing enhanced logistics and secure markets
- Strong sense of ownership and commitment of the implementing agencies resulting from close involvement of communities and cooperation between cities early in project design
- Responsibility delegated to the agencies operating at the local level, high staff continuity, and a solid network of extension officers

Source: World Bank project database.

By 2001, the total area and production of food fish under rice-fish culture reached 1.5 million ha and 0.85 million tons, respectively.[22] With intensified management and increased input, including artificial feeding, rice farmers could gain an average net profit of $1,813/ha/yr from the production of high-value species such as giant freshwater prawn and Chinese mitten-handed crab (Xiuzhen 2003). Improved technologies borrowed from pond aquaculture and growing market demand for fish led to adaptive management of rice-fish-farming practices, including changes in the choice of species and scale of operations.

In less-developed remote regions, financial support in the initial stage was a key factor to help the resource-poor farmers in paddy field renovation and startup operations.[23] Tax exemption is applied to rice-fish culture in places where it is promoted for poverty-reduction purposes. The main drivers for aquaculture in rural development in China included a set of liberal policies, such as liberal land use and farm management, which created a household responsibility system for diversified operation of land and capital. Liberalization of price controls on products and inputs provided the economic incentives to adopt aquaculture on rural farms (Wang 2001). Government invest-ment in aquaculture research and development (R&D) benefited farmers by prioritizing and diversifying species combinations and improving farm man-agement, leading to huge growth in pond productivity from 750 kg/ha in 1980 to 4,500 kg/ha in 1998.

Despite the huge leap in productivity growth, China's aquaculture has shown declining profitability because of increased production costs, declining genetic quality, and higher incidence of diseases. To improve productivity and sustainability, government policies encourage vertical integration of aqua-culture production in fish-growing areas, while supporting diversification of species, disease control, development of standards, and improved quality of aquaculture products.

India

Rice-fish farms are a low-cost alternative for poor farmers in India. India has rice-fish farms covering 2 million hectares, the largest reported area for such production in a single country (Halwart and Gupta 2004). The practice cuts across different ecosystems, from the terraced rice fields in the hilly terrain in the north to *pokhali* plots and deepwater rice fields on the coasts. In between are the mountain valley plots of northeastern India and the rain-fed or irri-gated lowland rice fields scattered all over the country. The species grown are just as diverse, with more than 30 species of finfish and some 16 species of shrimps listed. Most of the noncarp species and penaeid shrimp species are from natural stocks entering the rice fields with the floodwaters. Production rates vary from 3 kg/ha/yr in deepwater rice plots relying on natural stock of mixed species to more than 2 tons/ha/yr of tiger shrimps (*Penaeus monodon*) in shallow brackish-water rice fields.

Opportunities and challenges for poor people in community-based aqua-culture projects. Fish farming in leased public water bodies through commu-nity-based approaches has given a promising alternative to poor farmers in hundreds of villages in Orissa, Jarkhand, and West Bengal (STREAM 2005). Despite the constraints (see box A4.2), by applying pond polyculture technol-ogy through credit and extension support, substantial additional income is earned by members of groups organized under community-based fish farming.

Indonesia

Integrated livestock-fish farming in rice-based system is profitable, yet uptake is low. Integrated livestock-fish farming provides a viable option to poor rice farmers in Indonesia, earning them a higher net income than an aver-age government officer. Combining chicken raising with fish culture in earthen ponds on rice farms allowed farmers to optimize the utilization of on-farm wastes and supplement feed and fertilizer inputs to increase farm production and net household income (box A4.3). Although this is a relatively low-cost agribusiness, a majority of poor farmers find it difficult to adopt the technology because of the lack of credit and the high initial cost of digging ponds on the farm.

Large shrimp farming in tambaks transforms Indonesia's coastal villages. Traditional coastal rice areas (or *tambaks*) are being consolidated for shrimp production, which has a 90-day production cycle. New owners/leaseholders of *tambaks* are often urban investors attracted by shrimp farming's profitability. The operations and management of shrimp farms are usually performed by locals and financed by informal lending or investment. Large operation of *tambaks* for shrimp farming has attracted other forms of investment in rural areas, such as electricity, roads, and water, connecting remote rural hinterlands with urban commercial areas. The transformation has provided former farmers with an alternative livelihood as caretakers, managers, or farm workers. Financial institutions have shown little interest in providing capital to local people who want to invest in shrimp farming on lands they possess or hold under lease rights. The absence of clear land titles is a hindrance in securing institutional financing.

Batu Kumbung is a small village in Lombok, one of the major fish-producing areas in Indonesia. Many farmers here practice traditional rice–freshwater fish farming, where fish are kept in selected, small rice plots. The fish are fed daily with leftover rice from the household and are harvested before the rice. At the onset of the next rice season, farmers obtain new fish seedstock at no charge from well-off villagers who have their own hatcheries.

To sustain fish production and enjoy continuous income throughout the year, farmers began adopting integrated livestock-fish-farming systems, often with poultry, gouramy, carp, and a species of tilapia, locally called *mujair*. After 25 years of integrated livestock-fish farming, the farmers believed that the system offers significant benefits (profits). One farmer estimates the monthly net income from fish to be about $150, while the income from fish in the traditional system is about $50 per rice production cycle. The additional income from poultry is also significant. Although it is necessary to purchase feed and medicine, the poultry offer a very attractive monthly net income of $120. The total net income from integrated livestock-fish farming is almost twice the average monthly salary of a government officer ($150). The family not only enjoys a proper standard of living, but also has easy access to good, protein-rich food, such as fish, meat, and eggs.

Source: Juniati 2005.

Vietnam

The majority of Vietnam's rural people are still engaged in farming. Traditional practices—such as rice-fish and rice-shrimp culture, and integrated fish, livestock, and crop cultivation, including the widely known VAC system—provided an entry point to improve income and livelihoods within the limits of available land resources before moving toward intensive commercial aquaculture supported by more liberal land use policy and the opening up of export markets. However, even in some of the more advanced aquaculture practices, such as *basa* (catfish aquaculture) and shrimp farming in the Mekong delta, or intensive tilapia farming in the freshwater ponds in central and northern Vietnam, there are significant income and employment opportunities for poor people. The poverty-reduction dimensions of emerging large and intensive aquaculture of catfish and giant freshwater prawn farming are already visible from the participation of rural poor and women in the upstream and downstream supporting industries.

The success in seed production of the giant freshwater prawn and *basa* catfish provided new opportunities for farming systems with high economic performance, such as culture in rice fields (either integrated or alternative) and in ponds and garden canals (Wilder and Phuong 2002). *Basa* catfish is mostly reared in cages in the Mekong River, and its production in ponds together with *tra* catfish has been growing. There are 83 fish hatcheries and 32 giant freshwater prawn hatcheries in the Mekong delta area, and five processing and export factories operate in An Giang, the main catfish-producing province in Vietnam. These facilities provide significant rural employment opportunities.

The VAC farming system improved income and livelihoods in the early days of economic liberalization. The VAC, which is a totally family-managed farming system, can be found in irrigated lowlands, rain-fed uplands, and peri-urban areas of Vietnam. The system is a mix of annual and perennial crops, including fruits and vegetables, cattle, pigs, and poultry, with several species of Chinese and Indian carps grown in ponds. Annual yields of 2–3 tons/ha are commonly achieved while semi-intensive systems, especially with tilapia, may reach 4.5–5 tons/ha (Luu 1992). Since 1989, the Vietnamese government has distributed land for farmers and encouraged the development of the family economy not only by growing rice but also through diversified agriculture. In many Red River delta communities, VAC farming constitutes 50–70 percent of farmers' annual income, which can be three to five times higher than that from growing two rice crops per year (VACVINA 1995). The system is labor intensive and affords productive employment for people of all ages. The system also helps protect the production environment and improve family health and nutrition. Today, in Vietnam, the VAC system is considered to be an effective solution for poverty alleviation, dietary improvement, and prevention of malnutrition (Le Thi Hop 2003).

Aquaculture in the Mekong delta revolutionizing rural farms—employment and wage benefits. The Mekong delta is now home to Vietnam's intensive aquaculture, accounting for 85 percent of national aquaculture production. Intensive catfish (*Pangasius* spp.) culture in the Mekong delta started in cage culture in the 1960s and has occurred in ponds since 1999. These species are mostly destined for export to international markets. Cage and pond culture of catfish provides employment for over 11,000 households who operate their own aquafarms. Considering that each household hired two laborers to work in fish feeding, about 30,000 poor landless people were estimated to be working in catfish farming. On average, each hired laborer working on fish cages and ponds earns approximately $36–40 per month, or less than $2 per day. In 2003, there were also 5,300 workers with a salary income of less than $2 per day in five catfish export-processing factories based in An Giang province. About 3,000 people work in fish processing in Dong Thap, Vinh Long, and other parts of the Mekong delta. Poor women make up a high proportion of workers (more than 70 percent) in the processing factories. Several thousand people are also employed in related services sectors (for example, finance and credit organizations, fish feed and seed producers, and traders, veterinary services, storage, and transportation). The antidumping case (box 10) and the resulting measures brought by the United States led to the loss of employment among small-scale catfish-farming households, laborers, and people working in supporting industries (Bostock, Greenhalgh, and Kleih 2004).

Bangladesh

The potential of aquaculture for poverty reduction has attracted the attention of policy makers in Bangladesh (Bangladesh Planning Commission 2005). A strikingly high proportion of rural households (73 percent) are already engaged in some form of freshwater aquaculture (Mazid 1999). Considering the role aquaculture has played in providing employment and improving the livelihood of the rural poor (see table A4.1), a much greater contribution from aquaculture can be obtained if future government policies encourage improvement of productivity, expansion of aquaculture, and implementation of a workable strategy to encourage further gainful participation of the poor. At current market prices, aquaculture provides a more lucrative use of land than alternative activities; for example, a hectare of land devoted to aquaculture (carp) would generate at least 43 percent higher income for all factors engaged directly or indirectly in fish production than would a hectare of land under crop cultivation (Talukder 2004).

Two aquaculture models contributed to the success of aquaculture in targeting the poor in Bangladesh: (1) the provision of access to technology and training to directly transfer the benefits of technology to smallholders; and (2) the promotion of group- and community-based aquaculture. A wide range of technologies and farming models have proved relevant to the poor, including pond fish culture, rice-cum-fish culture, enhanced inland fisheries, floodplain aquaculture, and brackish- and freshwater shrimp culture. The economic benefits have accrued through direct fish production, employment on farms and in processing, service industries, enhanced food supplies, and on-farm fish consumption.

Pond fish culture is increasing local demand for labor. Over the past years, many underused water bodies and low-lying areas have been turned into private fishponds or community-based aquaculture. Even the upland rice fields are being converted into fishponds. All these changes suggest that fish culture is a relatively profitable activity. Based on a survey of 366 fish farmers, the average benefit-cost ratio in four major fish-growing districts (Mymensingh, Comilla, Bogra, and Jessore) showed that per hectare gross return from pond fish culture was taka (Tk) 115,788, representing a 1.89 return per taka of investment (Karim, Ahmed, Talukder, Taslim, and Rahman 2006). The major cost items in pond fish culture are human labor, fish seed, feed, and land use charges. Total labor requirement for culture of a 1-ha pond was 247 person-days for a one-year cycle of culture. Labor cost constituted about 30 percent of the total cost, of which 57 percent was hired labor.

Rice-cum-fish culture and floodplain aquaculture is raising production and farm income. Although cultivation of fish in irrigated rice fields has been

CHANGING THE FACE OF THE WATERS

Table A4.1 Annual Income by Stakeholder Group within the Bangladesh Shrimp Industry

| Stakeholders | Total household income (100%) | Household Income ($/year) | | Income Ratio (farm laborers = 1) |
		Shrimp-related activities (%)	Other farm and nonfarm activities (%)	
Hatchery owners	18,137	85	15	19.1
Shrimp farmers	15,150	78	22	16.0
Depot owners	4,729	47	53	5.0
Land leasers	2,445	24	76	2.6
Processing plant workers	1,752	51	49	1.8
Hatchery workers	1,529	71	29	1.6
Shrimp traders (faria)	1,309	66	34	1.4
Feed mill workers	1,255	78	22	1.3
Shrimp farm laborers	948	75	25	1.0
Shrimp seed collectors	634	38	62	0.7

Source: Islam et al. 2003.

widely practiced by rural farmers in Bangladesh for several decades, one of the earliest efforts in rice-fish farming was made by the Noakhali Rural Development Program in 1989. Through low-cost rural experiments, the program increased the production of cultured fish from 223 kg/ha to 700 kg/ha in 50 fields planted with local rice varieties. Results from comparative research between double-cropped rice culture and year-round fish culture showed that the net return for fish culture was more than four times the net return from double-cropped rice culture (Anik 2003; Rahim 2004; Talukder 2004; Akteruzzaman 2005; Karim et al. 2006). Today, rice and carp polyculture as well as carp with *galda* (freshwater prawn, *Macrobrachium* sp.) culture in rice fields are gaining popularity among farmers in the country (Mandal et al. 2004), contributing to on- and off-farm employment, income generation, and export earnings.

Direct benefits to poor and smallholders through access to technology and training. Since the early 1980s, a number of projects and programs that provided targeted extension and training services with or without credit facilities have had a major influence in accelerating technology adoption by smallholder and marginal farmers, and by increasing on-farm production, income and consumption rose several fold. Examples of projects include the Department of Fisheries (DOF)-DANIDA Greater Mymensingh Aquaculture Extension Project; DOF-DFID Northwest Fisheries Extension Project; Government of Bangladesh-ICLARM-IFAD Fish Culture Extension Evaluation Project; and WFC-USAID-DSAP Project (box A4.4) (Griffiths n.d.; Lewis, Wood, and Gregory 1996; Edwards 2000; Mandal et al. 2004; Thompson, Sultana, and Khan 2005). The World Bank and Asian Development Bank, the two largest multilateral donors in the country, have implemented fisheries and aquaculture projects to increase the participation of poor people through community-based and nongovernmental organization (NGO) mechanisms.

The direct beneficiaries of conventional aquaculture technologies and targeted extension services in Bangladesh have largely been small (0.5–1 ha) and medium (1–2 ha) landholders rather than the landless, or "functionally landless" (less than 0.2 ha) households (Edwards 2005). A significant proportion of small (34 percent) and medium (25 percent) landholding households were below the poverty line in 1995/96, the time period of most of these projects. These two landless groups did not have access to land and water resources or to the capital for adopting polyculture-based semi-intensive aquaculture, unless they were directly targeted by the projects for credit support and extension services. Leasing of village ponds through NGO credit support provided the poor with an initial entry into fish farming. Over time, competition from rich entrepreneurs pushed up the lease value and reduced their profitability. As a result, the poor lost their competitive advantage and often abandoned the practice.

**WFC-USAID Development of Sustainable Aquaculture Project (DSAP),
Bangladesh, 2000–05,** aimed to increase the number of small enterprises
producing and supporting the production of freshwater aquaculture products and improve household income. DSAP disseminated low-cost improved
technology packages to 35,000 demonstration farmers and to an additional
175,000 farmers by the end of the project. The project was implemented
through 33 partner NGOs. Productivity and profitability of carp polyculture
increased sharply—from 936 kg/ha to 2,660 kg/ha. Average annual fish consumption increased from 59.8 kg per farm to about 79 kg. Aquaculture
increased household income by 15 percent for grant farmers and by 36 percent for nongrant farmers. The participatory extension approach of the project increased female participation from 24 percent in 2001 to almost 50 percent in 2003. Anecdotal evidence showed that the higher female participation
increased not only their household income, but also brought a sense of confidence and higher social status for the participating female farmers.

Source: Mandal et al. 2004.

The DOF-DFID Northwest Fisheries Extension Project (NFEP), 1989–2000,
customized interventions for the poor by (1) promoting a fish farming system
that relied primarily on stimulating the growth of natural feed (plankton) produced in the pond using organic and inorganic fertilizers; and (2) mobilizing
and training poor seed traders to disseminate information on suitable aquaculture practices. The project trained more than 1,200 seed traders in fish
farming. Each seed trader had contacts with 40 farmers on average, and about
60 percent of fish farmers purchased fish seed from NFEP-trained seed traders.
However, poor seed traders favored relatively wealthier farmers—who could
purchase seed in cash—and did not offer to sell seed on credit to marginal
fishpond operators. Thus, access to working capital, including credit, proved
crucial for poorer fishpond operators.

Source: World Bank 2006. Agriculture Investment Sourcebook (available at: http://
siteresources.worldbank.org/EXTAGISOU/Resources/Module6_Web.pdf).

**Promoting group- and community-based aquaculture can benefit poor
and women.** Community-based fish-farming projects have developed opportunities for the poor to benefit from aquaculture.[24] One of the first major
attempts to use technology in aquaculture to directly benefit the landless poor
was the Grameen Bank's Joyshagar Fisheries Project in 1986. The project developed a community-based model capable of mobilizing the landless poor to
grow fish in underused freshwater ponds. Groups of landless people were guar-

anteed secure access to state-owned ponds, allowing specific tenure rights. The main lesson learned was that access to resources and credit by the poor is a key to success. Group leasing arrangements for ponds have been identified as a primary access route for the poor to farm fish.

To date, several NGOs have developed projects to help the poor access land and water. Using social influence and financial support under the ADB-financed Command Area Development Project, NGOs were engaged to organize the poor—mainly women—into groups, provide them with access to ponds for fish farming through private lease arrangements, help the groups acquire skills in fish farming and marketing, and provide them with microfinance services, including microcredit and savings facilities (box A4.5).

The Philippines

Tilapia farming as a small business and a source of household staple. Freshwater tilapia (cage and nursery) farming generates employment opportunities for small operators, caretakers, laborers, and their households, particularly in the rural areas where employment opportunities are limited and labor supply is abundant. Backyard/small pond and cage farms rely mainly on family labor. An estimated 24,000 people in Pampanga and Nueva Ecija (Central Luzon), including tilapia workers and their household members, depend directly on tilapia pond farming for employment. Caretakers and salaried workers on small tilapia farms earned P2,000–3,000 per month. In addition, they sometimes received free food and 10 percent of net profits. Some large tilapia farm-

---

Box A4.5 Group-Based Aquaculture Models in Bangladesh

Under the ADB-financed Meghna-Dhanagoda Command Area Development Project, NGOs were engaged to organize the poor—mainly women—into groups, provide them with access to ponds for fish farming through private lease arrangements, help them acquire skills in fish farming and marketing, and provide them with microfinance services, including microcredit and savings facilities. In 2000–03, farming of carps in 175 ha of leased ponds, held by 2,590 marginal and landless (owning less than 0.2 ha) people (96 percent women) gave an average annual net return of Tk 55 million (nearly $1 million). Each member earned about Tk 8,000 in net income from fish farming. In addition, several of the villagers earned a net income of Tk 175 per day from fish selling, and each seller employed two laborers on average. Likewise, fish seed traders who distributed fish fry and fingerlings earned a daily net income of Tk 136 to 245.

Source: ADB 2005.

ers hired caretakers at P3,000 per month and gave them 15–20 percent of net profits. Thus, tilapia pond farming provided employment and income benefits to poorer workers who were not in a position to establish their own ponds.

Fish consumption increased significantly in farming households in Taal, Batangas. On average, tilapia was consumed four to five times a week by cage-farming households and at least four times a week by tilapia nursery households. The supply of tilapia from cage farming in Lake Taal has helped keep tilapia prices stable, making tilapia more affordable to lower-income consumers (ADB 2005).

NATIONAL POLICY INTERVENTIONS AND PRO-POOR AQUACULTURE

China

In China, the main focus of aquaculture policies is as follows:

- Guarantee the supply of fish products and hence, food security.
- Absorb and use surplus rural labor, encourage women and young people to be engaged in production activities, increase farmers' income, and assist in poverty alleviation.
- Improve environmental and social awareness of aquaculture.
- Increase export earnings.

To foster rapid development of sustainable aquaculture, the State Council in 1997 issued the "Directive Notice on the Approval and Implementation of the Instruction of the Ministry of Agriculture to Further Expedite the Development of the Fishery Sector" (Hishamunda and Subasinghe 2003). This notice reformed and liberalized aquaculture and set the following additional guidelines to further develop the sector:

- Productivity improvement or higher output per unit of input will be the major effort to further develop the aquaculture sector. This can be achieved by improving the technology employed, promoting the use of high-value species and adjusting species mix or choice of the species cultured on the basis of market conditions.
- Strong efforts will be exerted to step up the greater and wider cultivation of the under- or unused "three uncultivated lands" (that is, water surface, mudflat, and flooded land suitable or fit for aquaculture) to make full use of the "cultivable" aquatic lands.
- Aquaculture licenses for newly cultivated aquaculture lands can be given to local collectives or villages. These licenses can be distributed to producers on a "production-by-production" contract basis, rental basis or temporary lease transfer basis, and even by auction.

- The Chinese aquaculture industry should be further liberalized and market regulations should be enacted and enhanced to promote large industry development.
- Aquaculture and the other agricultural industries should be equally developed in all possible regions of the country. Ecologically sound or ecosystem-based agriculture systems like integrated fish–livestock–silkworm–mulberry bush farming systems or fish culture in paddy fields should be actively promoted.
- In regions where the incidence of poverty is high or where poverty is endemic, especially in the midwest region of China, that are rich in or well endowed with aquaculture resources, aquaculture should be adopted not only as the source and means of livelihood and food security, but also as the main engine to drive the economy of the region in the fight against poverty and malnutrition.
- Enterprises, either public or private sector businesses engaged in aquaculture in the newly cultivated lands from the "three uncultivated lands," or in education or research in aquaculture, such as in the production or development of aquatic seed and new species, should be exempted from the agricultural special product taxes as an incentive.
- Available fiscal support and other assistance from all levels of government for developing aquaculture should be kept and/or even increased, if necessary to promote aquaculture investments.
- Because aquaculture production bases in suburban areas are the main suppliers of aquatic products for the cities, the acquisition and requisition of land for aquaculture should be prioritized and conversely denied or disallowed, if aquaculture land is being taken out of aquaculture for other non-aquaculture use, unless such requisition is in the national interests or for the common good of all citizens in the area.
- More investments should be made in the production of quality seed and fish health management to promote sustainable aquaculture development and management.
- Aquaculture health management problems (in particular, the treatment of fish diseases) should be solved by the guiding principle of "prevention first, prevention and treatment combined" through the establishment of a network for fish disease protection at different levels throughout the whole country.

Valuable lessons can be learned from the Chinese experience, including the following:

- Aquaculture can be developed in a sustainable manner to generate employment, improve income and livelihoods of rural and urban populations, and produce food.
- The engine for an economically resilient and sustainable aquaculture is the government's will and determination to establish sound policies and recognize the market will determine demand for the products.

- Several factors will strengthen aquaculture and ensure its sustainability and contribution to overall economic growth. These include rational use of available factors of production, improvements in legal and regulatory framework, and scientific breakthroughs in production technologies (Hishamunda and Subasinghe 2003).

Vietnam

Vietnam's Sustainable Aquaculture for Poverty Alleviation (SAPA 2000) strategy recognized the need for raising awareness on aquaculture opportunities, improving participatory approaches, and encouraging institutional capacity. It also recognized the gap between the needs of farmers and the services offered by extension institutions and issues of access to markets and financial services by the rural poor. The SAPA strategy aims to achieve the following:

- Enhance the capacities of poor people in rural areas to improve livelihoods through awareness raising and improved aquatic resources management and aquaculture.
- Strengthen the capabilities of institutions, and particularly local institutions, to understand and support the objectives of poor people in inland and coastal communities who depend on, or could benefit from, aquaculture.
- Share environmentally sound, low-risk, low-cost aquaculture technologies and aquatic resources management practices.
- Develop national policy based on lessons and experience from local pilots and intersectoral collaboration on strategies for addressing poverty.

Bangladesh

Bangladesh's plans[25] and strategic fisheries goals explicitly refer to aquaculture's role in poverty alleviation:

- Increasing productivity in inland aquaculture and in inland capture fisheries
- Raising income of poor fishers
- Promoting rice-cum-fish culture
- Strengthening fisheries research and extension

Planners in Bangladesh recommend the development of NGO extension services to relieve pressure on government budgets and continued collaboration with the private sector in pro-poor endeavors, and specify the following:

- Public-private partnership should be formed through government design of concessions that call for bidders to provide services in situations in which the government will not otherwise provide for the poor.

- Private utilities should be effectively regulated to ensure that the poor get better access to services at fairer prices.
- Responsible corporate citizenship must be encouraged.
- Privatization and divestiture should be linked to poverty reduction.

THE REGIONAL FRAMEWORK FOR SCIENCE AND TECHNOLOGY TRANSFER IN ASIA

Technology transfer and capacity building in Asia must be seen in the context of the drivers of investment and development of science-based aquaculture.

Market-driven development. Market forces were the primary driver of expansion and modernization of aquaculture in the region. The traditional preference for fish, a high population density, and a long-standing tradition of aquaculture moderated the perception of risk and encouraged investment. This, in turn, drove development and its attendant acquisition and evolution of technology. Competition and the drive for greater efficiency have resulted in development and adoption of more efficient production and management technologies and have stimulated consolidation and integration in the production chain. The combined pressures of the NGO conservation lobbies and international trade have driven the sector to be more competitive and environmentally friendly and to resolve environment-related trade constraints. This has hastened the transfer and diffusion of technologies and capacity building in policy, regulatory mechanisms, extension, training, and farm management, as well as in processing and marketing.

Investment in aquaculture development. The private sector in Asia drove the early development of commercial aquaculture and continues to do so. It has played a major role in acquiring technologies and transferring species and their culture technology in the region, initiating culture trials of new species, and driving technology development for commercial aquaculture. Private investment in high-value, high-risk aquaculture, for example, for grouper, crab, or abalone, proceeded even in the absence of public support (Aquaculture Asia 1997).

Nonetheless, national and international development banks and donor agencies played an important role in financing aquaculture. Technology transfer and capacity building were usually embedded in loan projects. Total official development assistance to aquaculture research and development in 1988–95 was about $995 million (in 1997 value terms), or the average annual input was about $124 million per year. During the period, the Asian region received 65 percent of commitments (Shehadeh and Orzeszk 1997). National investments more than matched this amount over the same period (Pillay 2001). Development banks were consistently the main source of such external funding, accounting for 69 percent of the total and in 1995, because of a drop in bilateral and multilateral assistance, they provided 92 percent of funding. The investment during 1987–97 required to achieve the global increase in annual production of 200 percent (to more than 36 million tons, or an annual average increase of 17 percent) has been estimated at $75 billion in year 2000 dollars (NACA/FAO 2001). This investment came largely from private sector resources and loans, grants, and government subsidies that supported the expansion of production facilities, research, health management capacity, feed development and production, hatchery development, processing facilities market channels, education and training, and technical and other forms of assistance.

Investments since 1976 have focused sequentially on four closely related themes: (1) higher productivity and increased economic returns; (2) improved environmental performance; (3) enhanced livelihood opportunities and socially responsible farming; and (4) market access and trade. The thrust of technology development and transfer and capacity building was closely aligned with these themes.

Sustainability drivers. Boom and bust cycles in the development of the shrimp culture industry and attendant environmental and social fallout drove the development of environmental regulations, the adoption of more environmentally friendly production methods, and innovations in shrimp production management. The impact of FDI, particularly from Taiwan, China, during the mid- and late 1980s on the development of shrimp aquaculture in Southeast Asia is illustrative (see box A4.6).

The harmonization of international standards for the quality and safety of traded products, certification for origin, and the advent of ecolabeling and organic aquaculture products have accelerated the transfer of technologies needed to meet those standards. Implementation of a mix of regulatory and

Figure A4.2   Changing Fortunes—Shrimp Aquaculture Production by Selected Producers

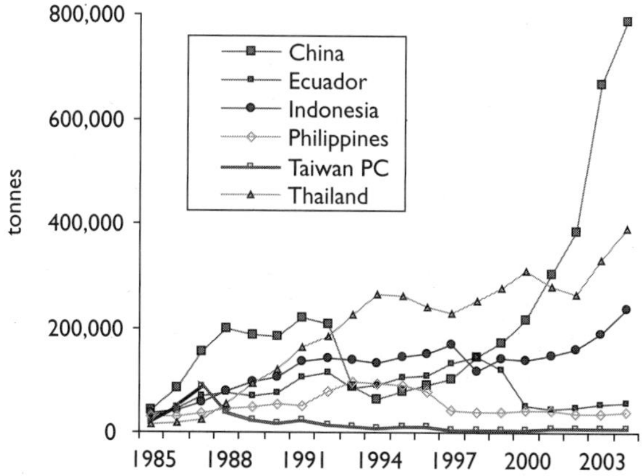

Source: FAO Fishstat 2005.

The advanced technology and expertise from the Taiwan (China) shrimp industry were transferred to Southeast Asian countries largely through joint ventures. Production costs were lower and as shrimp prices rose, profits soared. As a result of policy and regulatory lag or weak enforcement, the environmental costs were not internalized. As intensification grew, so did the environmental abuses, including the destruction of mangroves and discharge of untreated effluents at no cost to producers. Disease problems increased and, as regulations were increasingly enforced, costs rose and investment moved to another Southeast Asian country, leaving farms abandoned (see figure A4.2).

High profits attracted venture capitalists with a short investment horizon. Similarly, investment strategies of the small farmers also called for a single crop cycle investment horizon, partly in response to the high level of risk. The lesson brought home to governments, investors, and farmers was that the best insurance to investment is environmentally responsible farming. While the high-risk, high-profit phase of an infant industry may require venture capitalists, expansion requires effective environmental controls and mature investors who are committed to long-term sustainability.

Source: Fegan 1997.

voluntary management mechanisms jointly developed by government, industry, and farmer groups has led to the adoption of BMPs that address quality, safety, environmental, and social as well as ethical issues.

Development assistance. Multilateral and bilateral assistance agencies made substantive contributions to the development and strengthening of institutions and human resources and to technology transfer. Many bilateral agencies used their home countries' academic and research institutions to implement projects, so assistance was largely discipline or research focused. This led to the early development of human resources through training and graduate education, twinning of institutions, strengthening of university curricula, and support of research and research networks. Transfer of production technology came later and was associated with the more sophisticated intensive aquaculture systems and methodologies of developed countries. However, these earlier contributions to development of research capacity played a major role in the infusion of science into aquaculture. Bilateral donors also made significant contributions to the development of indigenous regional organizations and national research institutions. These contributions subsequently paid dividends by catalyzing the growth and evolution of the sector in the early to late 1990s. However, lack of interdonor coordination wasted resources, and the many projects taxed the resources of recipient countries and institutions.

The Kyoto Declaration—a turning point. The Kyoto Declaration on Aquaculture of 1976 marked a turning point in raising the profile of aquaculture and its recognition as a distinct sector with high potential and in attracting the support of donors and national governments (FAO 1976). The Kyoto Strategy set a target of a fivefold increase in aquaculture production in three decades (Bangkok Declaration and Strategy for Aquaculture Development Beyond 2000–2001), based on the following: (1) technical cooperation in the dissemination and use of known improved technology among developing countries; and (2) the development of new technologies. The Aquaculture Development and Coordination Programme of FAO/UNDP (ADCP), established to assist in this task, adopted a network approach using regional centers linked to national lead centers to address the issues posed by a diversity of aquaculture systems and species, and simultaneously promoted regional cooperation and strengthened national institutions. NACA, the Network of Aquaculture Centers in Asia-Pacific, became the hub of the Asian network. Ten years after its establishment under the ADCP, it evolved into a regional intergovernmental body with total regional ownership.

The ADCP approach was more broadly based and industry oriented than the bilateral projects at the time, with an aim to increase productivity. It incorporated development of a regional research network, capacity building in aquaculture policy and development, training of trainers on advanced technology, development of university curricula and regional research agendas, establishment of government stations for technology transfer, and funding of pilot demonstration projects. In the two decades that followed, the demonstrable increase in national outputs drove policy makers to allocate additional R&D funds to the sector. The investment in R&D generated production and productivity gains and further funding for R&D. The ninefold increase in global aquaculture production achieved by 2003 (89 percent of which derived from Asia) exceeded the strategy's targeted fivefold increase. FAO's program of Technical Cooperation among Developing Countries later provided a cost-effective mechanism for technology transfer from more advanced developing countries to others in the same region and elsewhere.

A diverse suite of initiatives and institutional arrangements at national, regional, and international levels evolved in Asia to provide a fertile institutional terrain for aquaculture technology transfer and capacity building. Numerous regional organizations have contributed to Asian aquaculture development, including the following: APEC, Association of Southeast Asian Nations (ASEAN), Intergovernmental Organization for Marketing Information and Technical Advisory Services for Fishery Products in the Asia-Pacific Region (INFOFISH), Mekong River Commission (MRC), NACA, South Asian Association for Regional Cooperation (SAARC), Southeast Asia Fisheries Development Center (SEAFDEC), and the South Pacific Forum. The regional financial institutions, education and training institutions (such as the Asian Institute of Technology), farmers' federations, and agribusiness alliances have all contributed in various forms. SEAFDEC and NACA provide technical advice to inform the development agenda of ASEAN. The following sections describe several of these institutions and their experiences, with a particular focus on NACA.

INTERNATIONAL CENTERS AND PROGRAMS

CGIAR and WFC. CGIAR and WFC have provided extremely important technology transfer links between formal agricultural research and diffusion efforts among countries and between countries and the National Agricultural Research Systems. Probably the two most valuable contributions of the CGIAR centers to technology transfer and capacity building are (1) supporting the transfer of genetic material and (2) transforming research results into usable technologies. A key contribution to aquaculture development in Asia was the development of the GIFT program by WFC (ADB 2005).

International Network on Genetics in Aquaculture. INGA (www.world fishcenter.org/inga), which is hosted by the WFC, strengthened national research capacities for the application of genetics in aquaculture, fostered regional and international cooperation, and assisted the development of national fish-breeding programs with an emphasis on tilapias and carps. INGA has provided international partnerships and links for the dissemination, information exchange, and further development of GIFT, and has assisted its members and other countries in sharing GIFT and other farmed fish germplasm for research, use in national breeding programs, and dissemination to farmers. It facilitated the formation of national networks for genetics in aquaculture. Limited financial and human resources are now constraining INGA's operations. WFC currently supports INGA from its core funds.

International Foundation for Science. IFS (www.ifs.se) has helped build a mass of highly trained aquaculture scientists in the region by supporting young researchers to pursue and use their research effectively, bringing together young scientists in small and focused workshops, and enabling young scientists to take part in international science symposia. The program has helped to prevent brain-drain for as little as $10,000 to $12,000 per grantee to support young researchers who have returned to their home countries after completing postgraduate studies.

ASEAN-EC Aquaculture Development Coordinating Program. The AADCP operated from 1988 to 1994 and made a significant impact on capacity building (of scientific manpower) and technology transfer by "twinning" EC and ASEAN institutions. Although continuation of the twinning activities has been constrained by the cessation of external funding, alumni of the program continue to serve as focal points for subsequent EC-supported projects.

NACA AND ASSOCIATED INITIATIVES

NACA began operating as a project in August 1980 and became an independent organization in January 1990. Its rationale, based on the Kyoto Strategy, was that sharing resources and responsibilities among institutions (and countries) is a practical and cost-effective means for addressing the diverse problems arising from a diversity of species, farming systems, and environments, and varying levels of development that the countries of the vast Asia-Pacific region face in modernizing, expanding, and sustaining aquaculture. The networking (and sharing) approach was also in line with the policy of governments to promote regional self-reliance through technical cooperation.

As an indication of cost-efficiency, the total cash input over 11 years from donor and government funding to NACA was $9 million, including the four-year UNDP/FAO Seafarming Development Project managed by NACA, which terminated in 1991 (Kongkeo 2001). While not directly attributable to NACA, aquaculture growth in the target countries between 1988 and 1997 was 11 percent by quantity—from 13.4 to 34 million tons and 9 percent by value—from $19.3 billion to $42 billion.

When NACA became an independent intergovernmental body, it adopted a major change in operational strategy. It had to (1) become self-sustaining to finance core activities (such as technical advice, information exchange, and network coordination and administration); (2) generate revenues by providing services against payments and developing programs and projects for collaborative assistance; and (3) forge partnerships with other institutions. These measures made it possible for NACA to continue as a focal point for implementing multilaterally and bilaterally funded regional and national projects. As an independent organization, the total government core (obligatory) contribution to NACA from 1991 to 2005 was $4.42 million, which leveraged an additional $10.53 million from external and other noncore sources of funding.

The in-kind contribution of members has not been quantified, but can be illustrated: China, starting in 1992, took over and funded under its TCDC program the annual three-month training course on integrated fish farming (IFF) in the NACA Regional Lead Centre in Wuxi. The course intake is usually 40 people from Asia-Pacific, Latin America, Africa, the Middle East, and Eastern Europe, and over 25 years, nearly 1,000 personnel have been trained. Centers in Thailand, Indonesia, and India offer or host regular and periodic courses for personnel from government, industry, farmer associations, and NGOs. Their courses are partially supported by the governments. Regional projects require national coordinators and on a no-fee basis, the governments and universities provide the institutional focal point for these regional projects.

When requested, NACA also manages national projects, for example, in Vietnam (donor funded) and in India (government funded for shrimp health management for small farmers). The results of these national projects are shared among countries through NACA's networking and TCDC activities.

The experiences in India have informed work in Vietnam and Iran on shrimp health management. In turn, these have benefited from the results of Australian Centre for International Agricultural Research (ACIAR)-assisted projects (in which NACA is also involved) on shrimp diseases in Thailand and Indonesia. ACIAR (www.aciar.gov.au/) has embedded a research and capacity-building component in the project in India. External expert assistance is kept to a minimum—a cadre of young local professionals and technicians is trained to provide the technical assistance to the farmers. Capacity-building activities include the farmers' associations and the institutions providing farm services.

The capacity-building and technology development/transfer achievements through NACA's coordinating role and collaborative work program can be summarized as follows:

- NACA has catalyzed broad-based support from various donors, pooled the scarce resources of national governments into a considerable regional resource, and coordinated the application in well-targeted collaborative and participatory regional and subregional projects.
- In terms of results, the broad-based collaboration on specific programs has involved numerous institutions and multiplied benefits to the institutions, governments, and industry. Cooperation in areas of mutual interests has mustered resources, expertise, and institutional support for regional projects, promoting synergy, avoiding duplication of activities, and expanding the range of beneficiaries. NACA has generated support for major regional and national activities from bilateral, multilateral, and investment agencies. Since 1990, there have been more than 65 such collaborative projects, workshops, training, assessments, and information activities of regional, subregional, and national as well as interregional scope.
- From the capacity-building perspective, training of national personnel and upgrading of facilities has created a multiplier effect for various assistance programs. The improved regional and national capacities brokered by NACA (for example, trained people, more efficient operating and management systems, and upgraded facilities) have facilitated implementation of donor assistance programs. The multiplier effects included the following: (1) wider dissemination of results; (2) assurance of follow-up activities within governments, thus ensuring continuity of project-initiated activities in the NACA work program; (3) use of strengthened national institutions by various assistance programs; and (4) building of formal intergovernmental processes.
- Cooperation and commitment are the basic forces that bind and move NACA. However, contributions to the management and operation of the organization are also essential. This is achieved through an agreed institutionalized arrangement as provided in the NACA Agreement, to which member governments accede and abide.

STREAM Initiative. Under STREAM (Support to Regional Aquatic Resources Management), a DFID livelihoods and poverty-oriented project in Southeast Asia (the Aquatic Resources Management Program) was sustained, institutionalized, and expanded in coverage by being incorporated into the regional program of NACA (www.streaminitiative.org). The STREAM initiative was established in 2001 as a NACA primary program element with multi-agency collaboration. Essentially, STREAM brought to aquaculture a greater understanding and use of the livelihoods approach developed by DFID and others. The initiative piloted these approaches in a range of Southeast Asian countries. STREAM fostered capacity development in public institutions, farmer groups, and NGOs and installed a regional communications mechanism to support the livelihoods approach and share expertise in pro-poor livelihoods projects. Central and state governments are incorporating the livelihoods approach in national policies and antipoverty fisheries programs.

Consortium on Shrimp Aquaculture and the Environment. The consortium, whose membership consists of FAO, NACA, WB, WWF, and recently UNEP, demonstrates the effectiveness of having a commonly agreed program with a focal point (that is, NACA) that does not have to deal with institutional bureaucracies (www.enaca.org/shrimp). The basic achievement of the consortium was developing a body of knowledge from a comprehensive worldwide study of the policies and practices in shrimp aquaculture and transforming that information into a set of broadly acceptable principles and guidelines for responsible shrimp farming by all stakeholders. Each partner brings its own funding and other resources into the program; in many cases, the individual partners have, individually or collaboratively, developed projects funded by foundations and other source of grants. The process demonstrates the cost-effectiveness of a consortium arrangement, and its institutional arrangement and working mechanism illustrate the potential role of a consortium in technology transfer and capacity building.

OTHER CONTRIBUTING INSTITUTIONS

Southeast Asian Fisheries Development Centre (SEAFDEC). SEAFDEC's Aquaculture Department (AQD) focused on the domestication of milkfish and the farming of penaeid shrimp, and through assistance mostly from the International Development Research Center (IDRC) and JICA, scientific manpower and facilities were upgraded (www.seafdec.org.ph/index.php). Additionally, technology products and capacity building were shared with other SEAFDEC member countries under its training and information program. When AQD became the regional lead center of NACA in the Philippines, its R&D program was expanded to several other species, its capacities further upgraded, and its regional outreach program strengthened and widened to include other Asian countries.

Asian Institute of Technology. The AIT, located in Thailand, is one of the key institutions in capacity building and technology transfer and diffusion in the region. Under its Aquatic Resources System Program, it trained many Asians in aquatic sciences and had fellowship programs for Bangladesh, Cambodia, China, India, Lao People's Democratic Republic, Myanmar, Nepal, the Philippines, Sri Lanka, and Vietnam. It also introduced farming systems research and extension (FSR&E) into the fishery extension programs of the lower Mekong River basin countries. The technology and skills were transferred from countries such as the United Kingdom and Denmark through its instruction and research programs. Donor assistance of DFID, DANIDA, the Swedish International Development Cooperation Authority (Sida), and USAID, as well as JICA and CIDA, supported the fellowship, research, and pilot outreach program on FSR&E. Alumni have become facilitators of AIT-initiated technology development, transfer, and diffusion programs in their home countries.

Other networks and societies. Networks and societies have had mixed success. Some have encountered sustainability difficulties. The Asia-Pacific Marine Finfish Aquaculture Network is an example of an institutional and people network with a coordinated well-structured R&D program that is partly supported by private industry.

NGOs. Civil society organizations have been active in a number of countries in the region. The delivery of assistance through NGOs is a response to the perception that, in certain circumstances, they are better able to deliver services at the local level or when state-to-state initiatives cannot be carried out. There has been an increasing reliance on the civil society sector in informing governments of gaps, problems, and needs related to development assistance. The technical expertise of sector institutions can complement the broader-based, people-oriented competence of NGOs. In addition to their advocacy and social organizing roles in Cambodia, NGOs, such as the Partnership for Development in Kampuchea (PADEK) and SAO Cambodia Integrated Aquaculture on Low Expenditure (SCALE), have undertaken capacity development and technology transfer. The apparent success of NGOs is based largely on anecdotal evidence; there is a lack of an empirical assessment of their performance (Simpson 2006).

Supplementary Statistical Information

Table A5.1 Aquaculture Production by Trophic Level of Cultured Species, 2003

Trophic Level	Million Tons	Percent
Carnivorous finfish	3.98	7.3
Omnivorous/scavenging crustaceans	2.79	5.1
Omnivorous/herbivorous finfish	16.02	29.3
Filter-feeding fish	7.04	12.9
Filter-feeding mollusks	12.3	22.5
Total animal aquaculture	**42.13**	**77.1**
Photosynthetic aquatic plants	12.48	22.9
Total animal and plant aquaculture	**54.61**	**100.0**

Source: Tacon 2005.

Table A5.2   Aquaculture Export Earnings in Some Developing Countries of Asia, 2003

Country	Major Export Commodities	Value ($ million)
Bangladesh	Penaeid shrimps	288
Cambodia	Penaeid shrimps	33
China	Penaeid shrimps, mollusks, seaweeds, unspecified marine and freshwater fillets	2,450
India	Penaeid shrimps	800
Indonesia	Penaeid shrimps, live grouper	1,644
Myanmar	Penaeid shrimps, live grouper, crabs	317
Philippines	Penaeid shrimps, seaweeds, live grouper, milkfish	200
Thailand	Penaeid shrimps, tilapia	1,1600
Vietnam	Penaeid shrimps, catfish (*Pangasius* spp.)	1,555

Source: FAO 2003.

Table A5.3 Top 40 Aquaculture Producer Nations, 2004 (excluding aquatic plants)

2004 Rank in Production (rank in 2003)	Country	Production 2004 (tons)	Percent of 2004 World AQ Production	Cumulative % of 2004 World AQ Production	Value (1,000 $)	Percent of 2004 World AQ Value	Cumulative % 2004 World AQ Value
1 (1)	China	30,614,968	67.3	67.3	30,869,519	48.7	48.7
2 (2)	India	2,472,335	5.4	72.8	2,936,479	4.6	53.4
3 (5)	Vietnam	1,198,617	2.6	75.4	2,443,589	3.9	57.2
4 (4)	Thailand	1,172,866	2.6	78.0	1,586,626	2.5	59.7
5 (3)	Indonesia	1,045,051	2.3	80.3	1,993,240	3.1	62.9
6 (6)	Bangladesh	914,752	2.0	82.3	1,363,180	2.2	65.0
7 (7)	Japan	776,421	1.7	84.0	3,205,093	5.1	70.1
8 (9)	Chile	674,979	1.5	85.5	2,801,037	4.4	74.5
9 (8)	Norway	637,993	1.4	86.9	1,688,202	2.7	77.2
10 (10)	United States	606,549	1.3	88.2	907,004	1.4	78.6
11 (11)	Philippines	512,220	1.1	89.4	700,854	1.1	79.7
12 (12)	Egypt, Arab Rep. of	471,535	1.0	90.4	617,993	1.0	80.7
13 (13)	Korea, Rep. of	405,748	0.9	91.3	979,825	1.5	82.2
14 (17)	Myanmar	400,360	0.9	92.2	1,231,230	1.9	84.2
15 (15)	Spain	363,181	0.8	93.0	431,990	0.7	84.8
16 (14)	Taiwan, China	318,273	0.7	93.7	942,789	1.5	86.3
17 (16)	Brazil	269,699	0.6	94.3	965,628	1.5	87.9
18 (18)	France	243,870	0.5	94.8	655,107	1.0	88.9
19 (20)	United Kingdom	207,203	0.5	95.2	593,300	0.9	89.8
20 (21)	Malaysia	171,270	0.4	95.6	324,285	0.5	90.3
21 (22)	Canada	145,018	0.3	95.9	398,907	0.6	91.0
22 (19)	Italy	117,786	0.3	96.2	365,415	0.6	91.5
23 (23)	Russian Federation	109,802	0.2	96.4	301,730	0.5	92.0
24 (25)	Iran, Islamic Rep. of	104,330	0.2	96.7	316,944	0.5	92.5

(continued)

Table A5.3 (Continued)

2004 Rank in Production (rank in 2003)	Country	Production 2004 (tons)	Percent of 2004 World AQ Production	Cumulative % of 2004 World AQ Production	Value (1,000 $)	Percent of 2004 World AQ Value	Cumulative % 2004 World AQ Value
25 (24)	Greece	97,068	0.2	96.9	365,561	0.6	93.1
26 (28)	Turkey	94,010	0.2	97.1	396,144	0.6	93.7
27 (27)	New Zealand	92,219	0.2	97.3	165,889	0.3	94.0
28 (26)	Mexico	89,037	0.2	97.5	291,329	0.5	94.4
29 (30)	Netherlands	78,925	0.2	97.7	118,268	0.2	94.6
30 (33)	Lao People's Democratic Republic	64,900	0.1	97.8	129,800	0.2	94.8
31 (34)	Korea, Democratic People's Republic of	63,700	0.1	97.9	58,250	0.1	94.9
32 (32)	Ecuador	63,579	0.1	98.1	292,077	0.5	95.4
33 (36)	Colombia	60,072	0.1	98.2	277,036	0.4	95.8
34 (35)	Ireland	58,359	0.1	98.3	121,284	0.2	96.0
35 (29)	Germany	57,233	0.1	98.5	171,274	0.3	96.3
36 (40)	Nigeria	43,950	0.1	98.6	124,396	0.2	96.5
37 (38)	Denmark	42,252	0.1	98.7	129,724	0.2	96.7
38 (31)	Faeroe Islands	41,879	0.1	98.8	129,153	0.2	96.9
39 (37)	Australia	39,331	0.1	98.8	260,020	0.4	97.3
40 (39)	Poland	35,258	0.1	98.9	79,895	0.1	97.4
	Rest of the world	491,758	1.1	100.0	1,635,365	2.6	100.0
	World total	**45,468,356**	**100.0**	**100.0**	**63,356,429**	**100.0**	**100.0**

Sources: Personal communication, FAO 2006.
Note: Production in metric tons.

Table A5.4 Projections of Food Fish Demand (demand/output in million tons)

Forecasts and Forecast Dates	Price Assumption	By the Forecast Date		Calculated Requirement from Aquaculture by the Forecast Date[a]				
				Growing fisheries		Stagnating fisheries		
		Global Per Capita Consumption (kg/yr)	Food Fish Demand	Total Output	Growth Rate (%)	Total Output[b]	Growth Rate (%)	Average Annual Increase
IFPRI (2020)	Real and							
Baseline	relative price	17.1	130.0	53.6[e]	1.8	68.6	3.5	1.7
Lowest[c]	are flexible	14.2	108.0	41.2	0.4	46.6	1.4	0.6
Highest[d]		19.0	145.0	69.5[e]	3.2	83.6	4.6	2.4
Wijkstrom (2010)								
2050	Constant	17.8	121.1	51.3[e]	3.4	59.7	5.3	2.4
	Constant	30.4	270.9	177.9[e]	3.2	209.5	3.6	3.5
Ye (2030)	Constant	15.6	126.5	45.5[e]	0.6	65.1	2.0	1.0
	Constant	22.5	183.0	102.0[e]	3.5	121.6	4.2	2.9

Sources: Ye 1999; Delgado et al. 2003; Wijkstrom 2003.

Notes: Demand and output in metric tons. IFPRI = International Food Policy Research Institute.

a. From 2000; 35.6 metric tons, three-year average of aquaculture output.

b. Assumes zero growth in food fish from capture fisheries after 2001.

c. Assumes an ecological collapse of fisheries.

d. Assumes technological advances in aquaculture.

e. Assumes a growth of output of food fish from capture fisheries of 0.7 percent per year to the forecast date.

Table A5.5 Total Per Capita Food Fish Supply by Continent and Economic Grouping in 2001

	Total Food Fish Supply (million tons live weight equivalent)	Per Capita Food Fish Supply (kg/year)
World	100.2	16.3
World excluding China	67.9	13.9
Africa	6.3	7.8
South America	3.1	8.8
Asia (excluding China)	34.8	14.1
North and Central America	8.5	17.3
Europe	14.4	19.8
Oceania	0.7	23.0
China	32.3	25.6
Industrial countries	26.0	28.6
Economies in transition	4.7	11.4
LIFDCs (excluding China)	22.5	8.5
Developing countries (excluding LIFDCs)	14.9	14.8

Source: FAO 2004.
Note: LIFDC = low-income food-deficit country.

Figure A5.1   Real Production Costs and Sale Prices of Farmed Atlantic Salmon

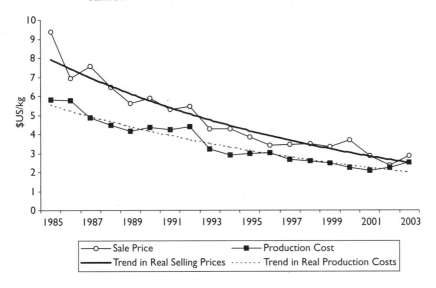

Source: LMC International Ltd. n.d.

Note: Projections of changes in fish and meat prices influence the supply models. In one model (Delgado et al. 2003), real fish prices are forecast to increase by 15 percent and prices of other animal substitutes by 20 percent. This is projected to reduce demand for fish products and increase substitution with lower-priced meat products, thus forecasting a slower rate of fish consumption to 2020 than over the last two decades. There is evidence of an increase in the real price of a significant proportion of fish for direct human consumption. The notable exceptions are cultured species, such as shrimp and salmon. In these cases, productivity gains are resulting in lower prices for cultured fish products, which are in turn affecting consumer preferences. This rising demand leads to further expansion in production, and competition drives prices even lower, expanding opportunities for further market penetration.

Figure A5.2   Aquaculture Production of Aquatic Animals by Main Species and Trophic Groups (% by weight)

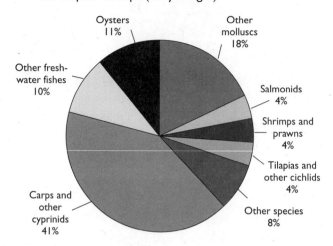

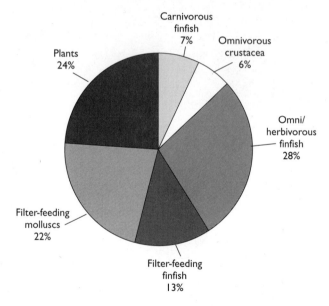

Source: FAO Fishstat 2005.

Figure A5.3   Aquaculture Production by Continent, 2004

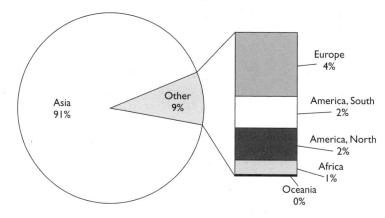

Source: FAO Fishstat 2005.

Figure A5.4   Growing Dominance of the Innovators: Global Atlantic Salmon Production

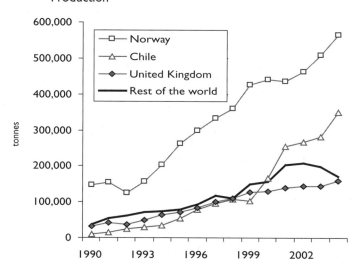

Source: FAO Fishstat 2005.

Table A5.6 Fish Consumption before and after Adoption of Improved Aquaculture in Bangladesh

Period	Production (kg/ha)	Percent of Home Consumption
Baseline	618	33
Postadoption		
Carp polyculture	2071	20
Tilapia monoculture	2208	67
Silver barb monoculture	1131	33

Source: Ahmed 2005.

Table A5.7 Production by Major African Aquaculture Producers

Country	Tons 1994	Tons 2004	Value ($000) 1994	Value ($000) 2004	Growth Rate % (quantity) 2002–2004	Growth Rate % (quantity) 1994–2004
Egypt, Arab Rep. of	56,603	471,535	103,432	617,993	25	733
Nigeria	15,030	43,950	40,065	124,396	43	192
Madagascar	3,295	8,743	6,637	35,215	−10	165
South Africa	4,729	6,012	8,501	32,410	8	27
Tanzania	3,150	6,011	836	1,250	−21	91
Uganda	179	5,539	157	6,107	13	2,994
Zambia	4,530	5,125	12,458	8,717	11	13
Congo, Dem. Rep. of	650	2,965	715	7,419	0	356
Zimbabwe	130	2,955	523	6,205	34	2,173
Tunisia	1,137	2,524	7,548	14,287	28	122
Morocco	1,463	1,718	11,014	5,887	3	17
Sudan	200	1,600	400	2,280	0	700
Togo	150	1,525	270	2,480	49	917
Seychelles	164	1,175	2,132	8,331	402	616

Source: FAO Fishstat 2005.

Figure A5.5　Global Aquaculture Production

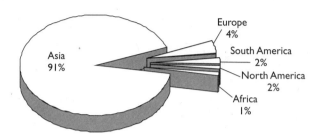

Africa >1% of global production

Source: FAO Fishstat 2005.

The Diversity of Aquaculture Production Systems and Business Models

SELECTED MODELS

Irrigation Systems

Integration of aquaculture with irrigated farming systems is an important means to enhance water productivity and overall environmental sustainability in Israel (see figure A6.1), Australia, Bangladesh, China, Indonesia, and elsewhere (FAO/NACA/CIFA 2001). Cage culture in irrigation canals and reservoirs is o+ ne approach; another is to lease sections of irrigation canal. Grass carp are widely used to keep canals free of weeds.

Investing in Common Property—Stock Enhancement

Because most wild fish resources are common property, many stock enhancement programs are managed and financed from public funds. In other cases, hydropower corporations may have an obligation to reseed dammed rivers. Alaska has an interesting model in which private nonprofit corporations owned by harvesters manage the salmon restocking (Smoker 2004). Public hatcheries have been transferred to the corporations, which derive revenues from a tax on salmon catches and from sales of a proportion (typically 40 percent) of the seed.

Figure A6.1   Schematic Diagram of an Integrated Commercial Farm in Israel

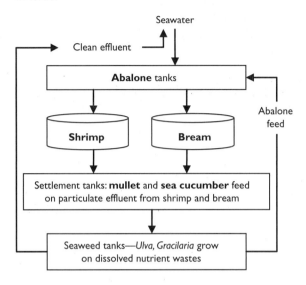

Source: Adapted from Davenport et al. 2003.

Recirculating Aquaculture Systems (RAS)

In the Netherlands, RAS has been a standard commercial practice for some 15 years, farming eel, catfish, tilapia, and turbot, and more recently, sole. RAS can reduce water needs from several cubic meters to less than 100 liters per kilogram of production in flow-through systems. Similarly, the chemical oxygen demand (COD) discharge may be reduced to 10 percent (equivalent to a stagnant fishpond). In Europe, environmental regulations have encouraged the development of RAS, especially when taxes may be levied on COD, nitrogen, or phosphate discharges.

Low Water Discharge Systems

A range of commercial, intensive, low-discharge systems has been developed and successfully applied, particularly in countries where there are serious water shortages or where water carries the threat of endemic parasites and diseases (Phillips, Beveridgand, and Clarke 1991). These systems include (1) combined intensive-extensive systems based on recycling water between intensive ponds and a reservoir used as a water-treatment component; (2) closed recycling systems in which water is cycled through devices to filter out suspended matter and to biodegrade organic components, ammonia, and nitrite; (3) "greenwater" closed recycling systems that combine stripping of nutrients by microalgae with the removal of suspended particles; and (4) low-exchange intensive ponds with a daily water exchange rate of up to 20 percent, with biological water treatment taking place in aerated, mixed water, similar to many industrial bioreactors, and nitrogen control through feed composition adjustment.

FOOTPRINTS OF PRODUCTION SYSTEMS

The ecological footprint is the amount of land, water, and resources required to sustain the aquaculture activity, including its waste disposal. Table A6.1 gives an example of the relative footprints of different culture systems using surface area as the common denominator; figure A6.2 contrasts generic characteristics of high- and low-trophic-level culture systems. In table A6.1, the fish content of pelleted feed is converted to area based on the notion of a "supporting marine production area."

Ecological footprint estimates incorporating water volume and circulation are likely to produce quite different results. Plant-based fish feeds now achieve similar production, feed conversion, and survival; as fish-based diets and experimental results are scaled up, the replacement of fish meal and fish oil in the diet will considerably alter the footprint of intensive farming of carnivores. In South Australia, this footprint concept is integrated into "a triple bottom line" of economic, social, and environmental measures required under the Aquaculture Act of 2001, which has as its primary objective "to promote ecologically sustainable development of marine and inland aquaculture."

AQUACULTURE PRODUCTION SYSTEMS IN AFRICA

The characteristics of the more important fish culture systems used in Africa are described below.

Table A6.1 Ecological Footprints of Aquaculture Systems

Activity	Area of Ecosystem Support Required per Unit Farm Area (multiples of the farm area)	
	Production	Waste Assimilation
Salmon cage farming, Sweden	40,000–50,000	—
Tilapia cage farming, Zimbabwe	10,000	115–275
Salmon tank system, Chile	—	16–180
Shrimp farming (semi-intensive), Colombia	34–187[a]	—
Shrimp farming (semi-intensive), Asia	—	2–22
Mussel rearing, Sweden	20	—
Tilapia pond farming (semi-intensive), Zimbabwe	0–1	0

Source: Folke et al. 1998.

Figure A6.2　Generic Representation of a Range of Low- and High-Trophic-Level Aquaculture Production Systems

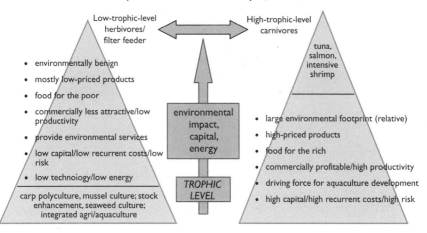

Source: Author.

Subsistence Ponds

Subsistence or smallholder ponds dominate freshwater aquaculture. In the 1990s, more than 90 percent of Sub-Saharan Africa's freshwater fish culture was based on earthen ponds, generally less than 500 square meters (m²) in surface area, and constructed and operated with family labor (Satia, Satia, and Amin 1992; King 1993). These ponds produce between 300 and 1,000 kg/ha (15–50 kg per crop) on an annual harvest cycle, usually corresponding to fingerling availability, water supply, or local demand. This low-input/low-output form of aquaculture contributes to local food security and livelihoods. Without new approaches, however, possibly based on Asian experiences, subsistence ponds are unlikely to make a major contribution to increased production. If concentrated in suitable peri-urban areas, where organic wastes may be used as feeds and where there is ready market access, subsistence production may transit to a commercial scale. This occurred in Cameroon, where urban fish prices were 48 percent higher and ponds were 72 percent more productive per unit area than those in rural areas, and IRRs were in excess of 34 percent (Brummet, Pouomogne, and Gockowski 2005).

Enhancement or Restocking

Increasingly, the productivity of water bodies is being enhanced through restocking. In West Africa, for example, controlled stocking of small dams (2000–10,000 m²) with or without fertilization is being used to increase typical background productivity of normally no more than 100 kg/ha to between 600 and 2500 kg/ha/yr

(Halwart and van Dam 2006; Oswald, Copin, and Montferrer 1996). In the Lower Shire River Valley of Malawi, local communities stock temporary waterbodies, or *thamandas,* with fingerling tilapias and catfishes, producing an average of 600 kg/ha (range 300–1,575 kg/ha) in a two- to three-month growing season (Chikafumbwa, Katambalika, and Brummett 1998). In Niger and Burkina Faso, traditional reservoir management systems have evolved in the direction of restocking after the annual drying. The fingerlings of *O. niloticus, Labeo coubie,* and *C. gariepinus* are produced through artificial reproduction of adults captured at harvest and held over the dry season. This restocking increases productivity from 50–100 kg/ha/yr to more than 600 kg/ha/yr (Baijot, Moreau, and Bouda 1994).

Integrated Agriculture-Aquaculture (IAA)

Malawi provides an example of IAA. These farms produce almost six times the returns generated by the typical Malawian smallholder (Scholz and Chimatiro 1996). The integrated pond–vegetable garden is the economic engine on these farms, generating almost three times the annual net income from the staple maize crop and the homestead combined. The vegetable-fish component contributes, on average, 72 percent of annual cash income (Brummett and Noble 1995). On a per-unit-area basis, the vegetable garden/pond resource system generates almost $14 per 100 m² per year, compared with $1 and $2 for the maize crop and homestead, respectively. By retaining water on the land, ponds enabled farms to sustain their food production and balance their losses on seasonal crop lands during severe droughts from 1991 through 1995. For example, in the 1993–94 drought season, when only 60 percent of normal rain fell, the average net cash income of a study group of rain-fed IAA farms was 18 percent higher than non-IAA farmers in an area with some of Malawi's severest poverty (Brummett and Chikafumbwa, 1995). Existing technologies for IAA may need to be further adapted to African conditions (IIRR and ICLARM 2001; Swick and Cremer 2001).

Rice-Fish Culture

In the 1990s, Madagascar undertook the privatization of its government fish stations, ceding or leasing these to private farmers or farmer associations. The operators are known as *Producteurs Privé d'Alevins* (private fingerling producers, or PPAs). A decade after divestment, the PPAs and their hatcheries are generally functioning well. They have lost their local monopoly over a supply zone and compete openly with each other in a profitable seed market. The Madagascar program is now experiencing new challenges from possible in-breeding of carp, and the program is examining options to improve extension support at the farm level and coordinate the large number of stakeholders, including many community-level NGOs (FAO Regional Office for Africa 1999.). Efforts to establish sustainable private hatcheries or nurseries as viable small businesses are also ongoing in eastern Uganda. However, in many Sub-Saharan countries, government seed supply stations built in the 1970s and 1980s are defunct and reliable seed supply remains a common problem.

Seaweed Culture

Tanzania produces 7,000 tons of *Eucheuma* cultured by an estimated 20,000 small growers. Using technology and seedstock imported from the Philippines, the growers depend on two large exporting companies (King 1992). Production technology is relatively simple and environmentally friendly, based on algae seedlings attached to a network of monofilament lines anchored to wooden stakes on tidal flats. Under contractual arrangements with the buyers, producers can earn about twice the average income of an entry-level civil servant. However, producers, mostly women, are also price takers and subject to exploitation as indicated by a major market maker.

> As a potential seaweed supplier trying to find the best village to work in, you should be delighted to find a village populated by consumers with no or little livelihood options. In this case we call [seaweed] farming the livelihood of last resort. Today we find the most productive and consistent farmers from villages like these. . . . In these places it is too arid to farm or the soil is unsuitable and the reefs have been destroyed and fish stocks decimated. . . . Your ultimate goal is to make seaweed farming become a way of life for the villagers. This happens after five or so years. At this stage people don't think too much about price, they just farm because they have always farmed. Their children will follow them. (FMC Biopolymer [USA] cited by Bryceson 2002)

Industrial seaweed is a highly consolidated industry—a few global corporations dominate the processing and distribution of nonfood products. However, farming and primary processing can be widely dispersed in rural and poorer communities and new niche markets for innovative products are emerging.

Small and Medium Enterprises

Aquaculture SMEs are expanding in Nigeria where the large government-funded Aquaculture and Inland Fisheries Project has targeted technical assistance and capacity building at this segment since 2003.[26] At least 100 farms with an estimated 60,000 ha under water produce 25,000–30,000 tons per year, mostly of catfish, which are highly prized in the Nigerian market (Moehl 2003). Grouped under the Aquaculture Association of Southern Africa (www.aasa-aqua.co.za), SMEs in South Africa are also making progress with a range of production systems (including freshwater crayfish, abalone, and ornamental fish). With the traditional focus on subsistence aquaculture, SMEs probably have not received the attention they merit because they have not been perceived to represent "the poor."

Large Commercial Aquaculture

There are an increasing number and range of successful large commercial aquaculture investments in Africa, and the commercial subsector is now estimated to account for approximately 65 percent of the total freshwater and brackish-water production and almost 100 percent of the mariculture production (Hecht 2006). The most notable include shrimp farms in Madagascar and Mozambique (export); tilapia in Zambia (ponds, local markets); Zimbabwe (cages, export); and Ghana (cages, local markets). New investments are coming in Uganda (cages, tilapia, export) and Kenya (cages, tilapia, local markets). Most of these investments involve foreign capital and rely on foreign or foreign-trained technical expertise and target growing urban African markets or international trade. All are vertically integrated to one extent or another, including feed manufacture, fingerling production, selective breeding programs, processing plants, and local and export market arrangements. The import of feeds, technology, and services compounds the relatively high investment risk with higher costs (compared with Asia),[27] and exporters rely on preferential tariffs for profit margins. Supply of feeds, both pelleted and low-cost wastes for subsistence ponds, is a major limiting factor on aquaculture expansion throughout the region.

Market-Driven Aquaculture

In Africa and elsewhere, domestic and export markets are driving aquaculture production. In Malawi, fish prices have increased by more than 350 percent in the 1999–2003 period, while improving national sanitary control systems have increased market access to the European Union. Madagascar, Mozambique, and Nigeria provide other examples of market-driven production. Private commercial tilapia farms are showing promise in Zambia, Zimbabwe, Ghana, Nigeria, and Malawi (Roderick 2002). The reviews of past aquaculture in Africa and recent success stories suggest that creation of wealth through a market-driven aquaculture strategy is more effective than a strategy with a social objective, such as food security (Hecht 2006). It can be argued, however, that the target groups of past failures in aquaculture development in Sub-Saharan Africa (the landless, or subsistence farmers) are less disposed to adopt aquaculture than are entrepreneurs and SMEs—the target group for market-driven aquaculture.

THE CASE OF SHRIMP CULTURE IN MADAGASCAR

Starting with an FAO project in the 1990s, Madagascar now has seven shrimp farms producing more than 8,000 tons per year on 2,250 ha. By 2002, the industry had created more than 4,000 jobs, contributing 5.9 percent of fiscal receipts and 6.9 percent of GDP. By constituting large farms as "duty-free zones" in relatively remote areas, the investments created new communities complete with infrastructure and services. How was this achieved?

Madagascar has good natural conditions for shrimp aquaculture and already had a well-developed shrimp trawl fishery. The FAO project helped build capacity and a basic shrimp culture strategy, implant sound environmental practices, and develop legislation. FDI worked through joint ventures to build the initial farms. Many of these investors had links to the existing shrimp-fishing industry. The chosen farm sites were free from land and resource disputes.

For environmental reasons and by agreement with the IFC, which financed some farms, only 3 percent of the concession area was used for shrimp farming. This was monitored by satellite mapping. Initial production systems were largely semi-intensive (2–7 tons/ha/yr), which helps avoid disease, and stocking densities were linked to the carrying capacity of the site. In practice, this means that for each 1 ha of shrimp pond, 2–3 ha of mangrove is required to remove nitrogen and phosphates from the effluent. Only indigenous species are cultured and these disease-free strains grow to exceptionally large sizes and command the highest prices. Investors must prepare an EIA. Confidential annual farm audits help enforce a code of conduct and promote mitigation measures with respect to environmental measures that are required as a result of the EIA.

More recently, other production systems are being used: family/artisanal shrimp farms using extensive production (less than 300 kg/ha/yr), "industrial" intensive farms (more than 7 tons/ha/yr), and a nucleus estate/contract farming model (under consideration). An aquaculture master plan helps guide development, while policy checks help ensure that only qualified investors (that is, those with expertise, capital, access to markets, and demonstrated bona fides) are granted concessions. An annual economic assessment of the shrimp culture industry is undertaken to monitor its contribution to the nation's economic development. The assessment tracks such key indicators as net foreign exchange, employment, and fiscal receipts.

PRODUCTION SYSTEMS AND CULTURED SPECIES IN LATIN AMERICA

Subsistence Aquaculture

Although in principle aquaculture is an effective means of improving rural livelihoods, it has shown little progress and a trend toward abandonment. The activity remains dependent on public and international support. Production is done in private or communal ponds, with about 85 percent of subsistence farms being family enterprises. Extensive, culture systems are widely practiced, generally using natural (primary) production and agricultural by-products for feeds. As in Africa, this involves mainly freshwater culture (mostly tilapia in combination with silver and common carp, *pacu*, and catfish), with some involvement in coastal culture of mollusks (oysters) and macroalgae. The family is in charge of pond management and the culture of the fish, while technical advice is provided by government agencies, albeit infrequently. Products from this sector are consumed by the family or sold locally. In Chile, culture of macroalgae is managed primarily by private companies and unions. The unions include artisanal fishermen who produce and market their products within one group. Artisanal producers live at low economic, social, and cultural levels, conditions that do not allow access to technology, information, markets, or credit. This situation severely limits the artisanal producer from further development.

Restocking of Inland Waters

Information on restocking of natural waters and artificial impoundments (for sport fishing and capture fisheries) is scarce, but Cuba, Mexico, and Brazil are the leading producers of fingerlings for stocking. Species stocked include trout, carp, catfish, tilapia, and silverside. Starting in the 1960s, Cuba had developed the extensive culture of tilapia and carp in large and small water bodies. Hybrid catfish was introduced recently and stocked in reservoirs with good results. In the northeast of Brazil, traditional culture of native species has been developed in many small and medium reservoirs.

Intensive Industrial Aquaculture

These large enterprises are usually incorporated as share capital companies. Joint ventures are common. Characteristically, they use intensive production systems (cages, ponds, raceways, tanks), sophisticated technology, and high expenditure in infrastructure and operating costs. They have large-volume production of salmon, tilapia, and shrimp. They have better access to national and international credit than do SMEs. There has been considerable vertical integration in these companies, both in freshwater (tilapia) and marine farming (salmon, mollusks). In Chile, more than half the industry is integrated with companies that supply materials and services. The main objective in recent years has been for companies to optimize their production processes and increase productivity by adopting new technologies and management procedures.

SMEs

Semi-intensive culture systems, particularly in freshwater farming, are the usual choice for the semi-industrial producer group, which is also characterized by owner management. Technical direction and management advice is usually provided by outside specialists. The farmed species include tilapia, shrimp, trout, and mollusks.

Main Species Groups

Aquaculture production from the region has been expanding at an average annual rate of about 13 percent during 1994–2004. The main produced species by volume are salmonids, mainly Atlantic salmon (44 percent); shrimp (23 percent); freshwater fish, mostly tilapia (23 percent); and mollusks (22 percent). Atlantic and Pacific salmon are farmed exclusively in Chile. The three major species groups are showing healthy growth rates. Details of the produced species are as follows:

- **Salmon.** Atlantic and Pacific salmon are cultured exclusively in Chile. There are 60 commercial salmon farms and 1,400 hatcheries. Between 1994 and 2004, production increased more than fivefold, from 109,000 tons to 586,000 tons—an average annual growth rate of about 19 percent. Until recently, salmon and trout farming had depended on the importation of eyed eggs (from the United States, United Kingdom, Norway, and Ireland).
- **Shrimp.** Shrimp farming started in 1968 and was decimated by diseases during 1998–99, particularly in Ecuador. It has been associated with negative environmental impact in coastal areas. With the adoption of BMPs and biosecurity measures, production has recovered from 153,731 tons in 2000 to 289,330 tons in 2004 (average annual growth of 7 percent). There are commercial farms (SMEs and large enterprises) in 20 countries of the region, and the industry has generated 750,000 direct and indirect jobs in the region. Governments perceive the industry as an engine for diversification of the economy, job creation in depressed areas, and earning of foreign exchange. The top five producers are Ecuador, Mexico, Honduras, Panama, and Colombia.
- **Tilapia.** Tilapia culture increased rapidly from 29,040 tons to 145,904 tons during 1994–2004, with an average annual growth rate of 18 percent. Growth was driven by the diversification trend in some countries combined with increased demand in foreign and intraregional markets. More than 90 percent of tilapia is produced by six countries: Cuba, Colombia, Mexico, Brazil, Jamaica, and Costa Rica.
- **Mollusks.** Mollusks were first cultured in Chile in 1921, having been developed mostly in the northernmost and southernmost countries of Latin America. Chile accounts for the bulk of production; other producers are Brazil, Mexico, Cuba, and Peru.
- **Macroalgae.** Macroalgae, such as *Gracilaria* sp., are produced in extensive systems in estuarine or marine environments. Algae are produced in Chile, República Bolivariana de Venezuela, and Peru, while experiments are being carried out in Panama and in Brazil.
- **Endemic fish species.** The culture of such endemic species as *Colosomas* and *Piaractus* (characids) is more common in Brazil, Argentina, Colombia, Peru, República Bolivariana de Venezuela, and Uruguay. Cultivation is generally in ponds under semi-intensive conditions, often in polyculture with tilapias.

Guidelines for the Preparation and Implementation of Aquaculture Projects

INTRODUCTION

The following guidelines were formulated on the basis of experience from completed Bank aquaculture projects and on the most recently identified best practices in the field of aquaculture (adapted from Braga and Zweig 1998). The goal of **these guidelines** is to facilitate the evaluation, preparation, and implementation of future aquaculture projects.

The actions and procedures recommended in these guidelines relate to the preparation and identification or the implementation phases of the project cycle. *The recommendations are presented in italics.* Examples and technical explanations from Bank projects in support of each recommendation are available in the Review of Completed Bank Aquaculture Projects.

PROJECT PREPARATION AND IDENTIFICATION

The first and most important step in achieving a successful project is to identify local needs properly and design a project that will fulfill those needs in the most efficient way. To achieve this, it is necessary to consult with the local community and, whenever possible, involve it from the initial stages of the project cycle.

It is important to allow enough time to complete every required step of project identification and preparation; these steps include thorough water and soil quality analysis, assessment of possible environmental impacts, identification of market conditions, preparation of well-defined technical designs, and detailed studies of the institutional arrangement and local capacity available to implement the project.

Finally, the preparation team should define how the project would fit in with other development initiatives taking place in the project area and should resolve any possible real or perceived conflicts.

Recommended Procedure

Projects should be based on sound and simple concepts, and use an already successfully tested technology and institutional framework. Aquaculture projects must be designed on the basis of well-founded market analysis and technical information about the basic fish resources, environmental conditions and legislation, suitable technology, and prevailing marketing practices.

Projects often suffer from being too ambitious and complex for the existing situation, introducing several new concepts and technologies at one time, without the institutional and community support necessary to successfully implement such complex projects. Successful projects are more straightforward and based on concepts and institutions that already have some history of existence in the community.

Recommended Procedures

Demonstration projects should be modest, simple, closely monitored, and involve only proven technologies.

A pilot phase should be mandatory for projects introducing new technologies or credit and institutional arrangements.

All communities affected by the project should be involved in the preparation process from the beginning.

Project preparation and appraisal are fundamental stages for successful implementation; they should not be skipped or shortened, particularly for large projects covering diverse areas.

There is no substitute for devoting adequate resources and time to careful project preparation and appraisal, and the failure to do so carries heavy risks.[28]

Recommended Procedures

Projects must incorporate some degree of flexibility to be able to adjust to changes in market and environmental conditions, development of better technologies, and appearance of disease. Attention should be paid to possible shifts in government approaches, particularly during long preparation and implementation periods. As a rule, the more complex the project, the more difficult it is to respond to changing conditions.

It is often difficult or impossible to foresee all that can happen during or after a project's implementation. As a consequence, projects should be structured with some degree of flexibility to be able to adapt to unexpected events.

To keep as many options as possible open for the future, when preparing a project, the following questions should be explored:

- What is the impact on project management and sustainability in the event of changes in government approaches or market conditions?
- How would the project be affected by significant changes in foreign exchange rates or in prices of raw materials and services needed for project management?

Answering such risk scenario questions may help design a project that would be more resilient to changes in the prevailing conditions.

Socioeconomic Issues

Among their primary goals, most aquaculture projects include the production of high-quality protein for human consumption and an increase in income and economic activity in the community. Because human activity is deeply interrelated with these projects, all forms of interactions between the project and the local communities must be thoroughly investigated. This can be best done by involving local people from the initial stages of project identification and preparation; local communities should be consulted on their ideas about the project, how it might affect their lifestyles, and how the community sees itself interacting with it.

Recommended Procedure

A good understanding of the socioeconomic and environmental characteristics of the project area is fundamental, especially of the land ownership and titles in areas involving small landholders.

Socioeconomic issues must be thoroughly reviewed during project preparation, including fishing and eating habits of local populations, existing relationships among local groups, and opportunities for private involvement in the project.

Recommended Procedure

Private sector participation should not only be anticipated but also encouraged, and whenever possible it should be specified in the original project design.

Environmental Issues
Recommended Procedure

Possible environmental impacts, both positive and negative, must be thoroughly addressed in all phases of aquaculture projects.

Sustainability Issues
Recommended Procedure

Sustainability is highly correlated with overall project quality, including technical and institutional aspects, and more specifically with whether farmers have access to credit and good extension support to ensure adequate managerial and technical capabilities of pond operators.

Financial Issues
Recommended Procedures

The project needs to make sure that farmers will have access to credit.

Cost recovery mechanisms should be introduced in projects showing high profitability.

PROJECT DESIGN

Technical Aspects
Recommended Procedure

A primary requirement for a successful aquaculture project is a reliable supply of suitable quality water.

The usual sources for water in aquaculture projects are as follows:

- **Groundwater.** Generally more dependable and uniform than surface water and usually less polluted, although that is rapidly changing. Its major disadvantage is being devoid of oxygen, and it must be aerated before use; it may also contain toxic gases and high concentrations of certain dissolved ions and minerals.
- **Surface waters.** Should be carefully evaluated because they are subject to contamination, often carry high silt loads, and may contain wild fish, parasites, predators, or disease organisms. Will frequently require pumping, but the cost is usually less than pumping water from wells. Surface waters are often subject to environmental regulations that may change without much advance warning.
- **Alternative sources.** Include rainwater, saltwater wells, and water recycled after some other use, such as irrigation tailwater and cooling water from industrial processes.

Water needs vary greatly with the aquaculture system being used, and projects need different amounts of suitable water at specific stages of operation. Well water is the desired water supply for hatcheries; for culture ponds, economic considerations play an important role in choosing the water source and should not be ignored. Once water reaches the project area, one of two basic channel configurations is usually chosen to direct water in and out of ponds. The first uses a single channel for both functions; the other, which is more expensive, allows for separate inlet and outlet channels.

Recommended Procedure

The choice of site for project development must consider technical, managerial, marketing, and social constraints.

A range of variables must be considered when choosing the site for project development: proximity to water source and markets, topography and soil type, the landholding situation, and possible constraints resulting from the social structure and culture of local communities.

Recommended Procedure

Aquaculture projects should avoid dependence on wild caught seed supply.

Project performance can be hindered by the unreliability of wild seed supply.

Recommended Procedure

Projects must carefully plan and monitor the stocking densities used in culture ponds and lakes because high densities can be conducive to disease outbreaks.

Mortality caused by disease is one of the main problems encountered in the management of aquaculture schemes, especially in shrimp culture.

Recommended Procedure

The bidding process for equipment acquisition should give more weight to technical considerations and quality of equipment, and less to price.

Specifications and compatibility must be thoroughly checked for all proposed equipment; an exaggerated emphasis on low prices can contribute to the acquisition of low-quality or outdated equipment.

Recommended Procedure

When using private parties in project implementation, the selection should carefully consider both competence and experience in the type of work to be done.

Ensure that all hired parties have previous experience with the technology and the physical and social conditions prevailing in the project area.

Recommended Procedure

Plan for adequate time to complete civil works and contract awards, and allow for unexpected events.

After project implementation has started, changes in the original plans often prevent the engineering work from being completed within the allotted time. Careful project preparation, including pilot phase trials of new concepts and technologies, is essential to accurately predict the time required to complete civil works.

Institutional Aspects

The institutional framework for a project is as essential to its success as are economic, financial, and technical considerations.

Recommended Procedures

Keep project organization as simple as possible and clearly define the roles and responsibilities of people and agencies involved in the project, both existing and newly required. There must be a clear consensus among the Bank and all parties involved about the nature (public or private), objectives, organization, financing, and operating principles of the institutions involved.

It is highly recommended that, as much as possible, projects should work within the country's existing institutional framework; any eventual major innovation should first be tested in a pilot phase. Too many people involved in the decision-making process, as well as poor communication among the involved parties, can cause considerable delays in project implementation.

Recommended Procedure

Project agencies tend to work closer together at lower levels of government; placing responsibility at the district level rather than at the central level usually improves performance.

Agencies operating at the local level will usually have a better "one-on-one" relationship with the farmers than agencies operating at the central level, which can contribute significantly to project success.

Recommended Procedures

The agencies responsible for the technical aspects of the project should work closely with those responsible for giving out credit for project implementation.

Informing the local lending institutions about the technical and managerial aspects of aquaculture projects can significantly improve their willingness to provide credit for such initiatives.

PROJECT IMPLEMENTATION

Many problems can arise during project implementation, mostly because of insufficient preparation or a poor set of initial designs. Frequent Bank supervision missions that include the technical expertise needed by the project are essential for successful implementation; these missions can identify problems soon after, or even before, they appear, greatly reducing the time and cost of the implementation phase. When problems appear, stop and reevaluate.

Recommended Procedures

Thorough project preparation, including a good set of initial technical designs and detailed soil analysis, is the best insurance against the occurrence of technical problems during project implementation.

The most frequent technical problems encountered during project implementation were caused by one of the following factors: poor or incomplete initial designs, poor quality of engineering work, or unsuitability of soils for the proposed work.

Recommended Procedures

Appraisal and supervision missions should not be carried out with the same experts every time, although there should be at least one team member who will follow the project from beginning to end.

Some variety in the supervising group allows a project to be analyzed from different points of view; however, a degree of constancy is required to ensure continuity and a full understanding of all aspects of the project.

Recommended Procedures

Governments need to provide adequate incentives to keep staff and newly trained personnel in the project for long periods of time.
Borrowers should appoint and train enough staff as soon as the project is approved.
Good training of extension officers is essential.

A high degree of staff continuity during project implementation usually provides a strong sense of ownership and commitment from the implementing agencies and contributes to overall project success. It is common to underestimate the time required to appoint and train the staff needed to implement and manage aquaculture projects. This process should start as soon as the project is approved for implementation.

1. Consider the diversity of aquaculture systems. *Species:* An estimated 230 commercial species of finfish, mollusks, crustaceans, aquatic plants, turtles, and frogs out of a total of 442 cultured species (not counting ornamental fish), compared with 73 species in 1950. However, 25 species account for 90 percent of global production. *Aquaculture environments:* freshwater, brackish-water, marine; inland, coastal, and oceanic. *Production systems:* rain-fed ponds; irrigated or flow-through systems; tanks and raceways; sea ranching; closed recycling systems; pond-based recirculation; monoculture and polyculture systems; integrated livestock-fish farming; integrated agriculture-aquaculture. *Scale and intensity:* small, backyard ponds and hatcheries; industrial commercial operations; extensive, semi-intensive, and intensive.

2. There are some indications that Chinese production statistics for both capture fisheries and aquaculture may be overestimated.

3. Selected countries are as follows: (1) Asia: India, Indonesia, Japan, Thailand, Bangladesh, Vietnam, and the Philippines; (2) Africa: the Arab Republic of Egypt, Nigeria, and Madagascar; (3) Latin/South America: Chile, Brazil, Mexico, Ecuador, and Colombia; (4) North America: the United States and Canada; and (5) Europe: Norway, Spain, France, Italy, the United Kingdom, and Greece (Brugere and Ridler 2004).

4. Including ongoing and completed projects. Some projects had planned aquaculture components at the appraisal stage (an additional $194 million has been approved), but the investments did not occur, or outcomes of aquaculture components were not documented in Project Completion Reports (PCRs) or Implementation Completion Reports (ICRs).

5. Excluding a major study on Vietnam completed in 2005.

6. Removes three-quarters of total phosphorus and 96 percent of suspended solids.

7. See, for example, Article 9.3 of the FAO Code of Conduct for Responsible Fisheries (1995b).

8. Australia's Commonwealth Scientific and Industrial Research Organisation

(CSIRO) has begun research on a transgenic technology that creates functional sterility, so stocks can only complete their life cycle under culture conditions and any escapees are unable to breed or produce viable offspring.

9. See, for example, http://www.ifoam.org.

10. See various reports of the OIE Aquatic Animal Health Standards Commission.

11. The Network of Aquaculture Centres in Central and Eastern Europe (NACEE) is an interesting model. There is no agreement among governments but only among the participating R&D institutions (see http://www.agrowebcee.net/subnetwork/nacee/).

12. In Latin America, some initiatives have already taken place in this regard. See Aquatic Animal Pathogen and Quarantine Information System (AAPQIS) at http://www.aapqis.org/main/main.asp.

13. Projects include the Development of Sustainable Aquaculture Project (DSAP)-WorldFish; DANIDA-Mymenshingh Aquaculture; CBFM Projects; IFAD-Oxbow Lake; and the Care Cage Aquaculture Project.

14. Cambodia (PRSP), Oman (sixth Five-year Development Plan), and Malaysia (food security in a five-year plan) all make reference to aquaculture in national policy or planning documents.

15. Decision No. 224/1999/QD-TTG, December 1999 (Escober 2004).

16. Bangladesh Planning Commission 2005.

17. Fish exporters still face very high tariffs for selected species ("tariff peaks") and higher tariffs for processed or value added products ("tariff escalation") than raw materials.

18. According to WTO, antidumping initiations rose from 157 in 1995 to 364 in 2001 (down to 213 in 2004).

19. For example, Aqualma (Madagascar), Kafue Fish Farm (Zambia), and Lake Harvest (Zimbabwe).

20. The top producers in 2004 were Chile (52 percent), Brazil (20 percent), Mexico (7 percent), Ecuador (5 percent), and Colombia (4 percent).

21. The government established the first salmon farm in collaboration with Corporation de Fomento, a public development agency, and Fundacion Chile, a private organization created to facilitate technology transfer.

22. Compared with a total area of 0.85 million ha and food fish production of 207,000 tons in 1994.

23. Support from the government is usually delivered in the form of construction materials and seeds.

24. For example, the Grameen model (Watanabe 1993), or group-based fishpond lease-holding (for example, ADB's Meghna-Dhanagoda Command Area Development Project) or Daudkandi Community-Based Floodplain Aquaculture (Morshed 2004; Rahman et al. 2005) or DFID's Northwest Aquaculture Project (Lewis, Wood, and Gregory 1996).

25. See the Policy Matrix 4 on Agricultural Growth toward Poverty Reduction (Bangladesh Planning Commission 2005).

26. The AIFP was originally funded at $7 million, which was substantially reduced in 2005.

27. Recent work by the OECD's Sahel and West Africa Club indicates production costs for shrimp production in West Africa, which are well in excess of Asian costs.

28. *A Review of Completed Aquaculture Projects* (Braga and Zweig 1998) provides several examples of problems resulting from hurried preparation and appraisal phases.

REFERENCES

Ackefors, H., and M. Enell. 1994. "The Release of Nutrients and Organic Matter from Aquaculture Systems in Nordic Countries." *Journal of Applied Ichthyology* 10 (4): 225–41.

Aerni, P. 2001. "Aquatic Resources and Technology: Evolutionary, Legal and Developmental Aspects." Science, Technology, and Innovation Discussion Paper 13. Center for International Development, Cambridge, MA.

Aguilar-Manjarrez, J., and S. S. Nath. 1998. "A Strategic Reassessment of Fish Farming Potential in Africa." CIFA Technical Paper 32. Rome, FAO. 1998.

Ahmed, M. 2006. "Review of Pro-Poor Aquaculture Development in Asia." Unpublished report prepared for the World Bank.

———. 1997. "Fish for the Poor under a Rising Global Demand and Changing Fishery Regime." *NAGA Supplement* (July–December): 4–7.

———. 2004. "Outlook for Fish to 2020: A Win-Win for Oceans, Fisheries and the Poor?" In record of a conference conducted by the Australia Academy of Technological Sciences and Engineering (ATSE) Crawford Fund, "Fish, Aquaculture and Food Security: Sustaining Fish as a Food Supply," ed. A. G. Brown. Canberra, Australia: Crawford Fund. August 11, 2004.

———. 2005. "Trends and Prospects for Aquaculture in Developing Countries: Drivers of Demand and Supply in Changing Global Markets." PowerPoint presentation at the FAME workshop, University of Southern Denmark.

Ahmed, M. V., and M. H. Lorica. 2002. "Improving Developing Country Food Security through Aquaculture Development: Lessons from Asia." *Food Policy* 27: 125–41.

Akteruzzaman, M. 2005. "From Rice to Fish Culture: Process, Conflicts and Impacts." Unpublished. A report submitted to the WorldFish Center, South Asia Office, Dhaka, Bangladesh.

Allan, G. 2004. "Fish for Feed vs. Fish for Food." In record of a conference conducted by the ATSE Crawford Fund, "Fish, Aquaculture and Food Supply: Sustaining Fish as a

Food Supply," ed. A. G. Brown, 20–26. Crawford Fund: Canberra, Australia. August 11, 2004.

Anik, A. R. 2003. "Economic and Financial Profitability of Aromatic and Fine Rice Production in Bangladesh." M.S. thesis. Department of Agricultural Economics, Bangladesh Agricultural University, Mymensingh, Bangladesh.

Anonymous. 1997. Holmenkollen Guidelines for Sustainable Aquaculture. Oslo, Norway.

Aquaculture Asia. 1997. "The Rewards of Inefficiency." 2 (3): 1–2.

Aquaculture Association of Southern Africa Web site. 2006. Available at http://www.aasa-aqua.co.za (accessed 2006).

Argue, B. J., S. M. Arce, J. M. Lotz, and S. M. Moss. 2002. "Selective Breeding of Pacific White Shrimp (Litopenaeus vannamei) for Growth and Resistance to Taura Syndrome Virus." Aquaculture 204 (3): 447–60.

Asian Development Bank (ADB). 2005a. An Evaluation of Small-Scale Freshwater Rural Aquaculture Development for Poverty Reduction.

———. 2005b. An Impact Evaluation on the Development of Genetically Improved Farmed Tilapia and Their Dissemination in Selected Countries. ADB, Manila, October 2005. Available at http://www.adb.org/publications (accessed 2006).

Australian Centre for International Agricultural Research Web site. 2006. Available at http://www.aciar.gov.au/ (accessed 2006).

Baijot, E., J. Moreau, and S. Bouda, ed. 1994. "Aspects Hydrobiologiques et Piscicoles des retenues d'eau en Zone Soudano-Saheliene." Centre Technique de Cooperation Agricole et Rurale ACP/CEE, Wageningen, The Netherlands.

Bangkok Declaration and Strategy for Aquaculture Development Beyond 2000. 2001. In Aquaculture in the Third Millennium, ed. R. Subasinghe, P. Bueno, M. J. Phillips, C. Hough, S. E. McGladdery, and J. R. Arthur, Technical Proceedings of the Conference on Aquaculture in the Third Millennium, Bangkok, Thailand. 463–71. February 20–25, 2000. Bangkok: NACA; Rome: FAO.

Bangladesh Planning Commission. 2005. "Unlocking the Potential: National Strategy for Accelerated Poverty Reduction." GED, Government of People's Republic of Bangladesh. Planning Commission, Bangladesh. Unpublished.

Bartley, D. M., R. Subasinghe, and D. Coates. 1996. "Framework for the Responsible Use of Introduced Species." EIFAC /XIX/96/inf.8 Report of the Ninth Session of the European Inland Fisheries Advisory Commission, Dublin, Ireland.

Bisharat, N., and R. Raz. 1996. "Vibrio Infection in Israel Due to Changes in Fish Marketing." Lancet 348: 1585–86.

Bondad-Reantaso, M. G. 2004. "Development of National Strategy on Aquatic Animal Health Management in Asia." In Capacity and Awareness Building on Import Risk Analysis for Aquatic Animals, ed. J. R. Arthur and M. G. Bondad-Reantaso, 103–08. Proceedings of the Workshop held in Bangkok, Thailand, April 1–6, 2002, and Mazatlan, Mexico, August, 12–17, 2002. APEC FWG 01/2002; Bangkok: NACA.

Bondad-Reantaso, M. G., R. P. Subasinghe, J. R. Arthur, K. Ogawa, S. Chinabut, R. Adlard, T. Zilong, and M. Shariff. 2005. "Diseases and Health Management in Asian Aquaculture." Veterinary Parasitology 132 (3-4): 249–72.

Bostock, T., P. Greenhalgh, and U. Kleih. 2004. "Policy Research—Implications for Liberalisation of Fish Trade for Developing Countries." Synthesis Report. Chatham, UK: Natural Resources Institute.

Boyd, C. E., and W. Clay. 1998. "Shrimp Aquaculture and the Environment." Science America 58: 59–65.

Braga, M. I. J., and R. Zweig. 1998. "A Review of Completed Aquaculture Projects." Draft. World Bank, Washington, DC.

Brummett, R. E., and F. J. K. Chikafumbwa. 1995. "Management of Rainfed Aquaculture on Malawian Smallholdings." Paper presented at the PACON Conference on Sustainable Aquaculture, Honolulu, Hawaii. June 11–14. Pacific Congress on Marine Science and Technology, Honolulu, Hawaii.

Brummett, R. E., and R. P. Noble. 1995. "Farmer-Scientist Research Partnerships and Smallholder Integrated Aquaculture in Malawi." In *The Management of Integrated Freshwater Agro-Piscicultural Ecosystems in Tropical Areas,* ed. J.-J. Symoens and J-C. Micha. Wageningen, The Netherlands: Technical Centre for Agricultural and Rural Cooperation; Brussels, Belgium: Royal Academy of Overseas Sciences.

Brummett, R. E., V. Pouomogne, and J. Gockowski. 2005. "Development of Integrated Aquaculture-Agriculture Systems for Small-Scale Farmers in the Forest Margins of Cameroon." WorldFish Center, Yaoundé, Cameroon.

Buras, N. 1990. "Bacteriological Guidelines for Sewage-fed Fish Culture, in Wastewater-fed Aquaculture." In *Proceedings of the International Seminar on Wastewater Reclamation and Reuse for Aquaculture,* ed. P. Edwards and R. S. V. Pullin, 223–36. Calcutta, India, December 6–9, 1988. Bangkok: ENSIC, AIT.

Chesapeake Bay Net. 2006. Available at http://www.chesapeakebay.net (accessed 2006).

Chikafumbwa, F. J., K. L. Katambalika, and R. E. Brummett. 1998. "Community Managed Thamandas for Aquaculture in Malawi." *World Aquaculture* 29 (3): 54–59.

Chilaud, T. 1996. "The World Trade Organization Agreement on the Application of Sanitary and Phytosanitary Measures." Scientific and Technical Review—International Office of Epizootics 15: 733–41.

Cho, C. Y., J. D. Hynes, K. R. Wood, and K. Yoshida. 1994. "Development of High-Nutrient-Dense, Low-Pollution Diets and Prediction of Aquaculture Wastes Using Biological Approaches." *Aquaculture* 124 (1-4): 293–305.

Chobanian, E. 2006. "Aquaculture and Coastal Management Portfolio Review. Lessons Learned." Unpublished draft report. World Bank, Washington, DC.

Chopra, K., and P. Kumar in collaboration with P. Kapuria and N. A. Khan. 2005. "Trade, Rural Poverty and the Environment: A Study in Eastern India." Institute of Economic Growth, Delhi, India.

Convention on Biological Diversity. 1992. Available at http://www.biodiv.org/convention/articles.asp (accessed 2006).

Cullinan, C., and A. van Houtte. 1997. "Development of Regulatory Frameworks." In *Review of the State of World Aquaculture* 75–79. Rome: FAO.

Dallimore, J. 2004. "Traceability in Aquaculture." The INFOSAMAK Buyer-Seller Meeting. Edition 2004 and The First Value-Added Seafood Conference, Cairo, Egypt. April 26–28, 2004. Cairo: INFOSAMAK.

Davenport, J., G. Burnell, T. Cross, S. Culloty, S. S. Ekaratne, B. Furness, M. Mulcahy, and H. Thetmeyer. 2003. "Aquaculture: The Ecological Issues." The British Ecological Society Ecological Issues Series. Oxford, UK: Blackwell.

De, H. K., and G. S. Saha. 2005. "Community Based Aquaculture—Issues and Challenges." *Aquaculture Asia* X (4). October–December 2005.

De Silva, S. S. 2001. "A Global Perspective of Aquaculture in the New Millennium." In *Aquaculture in the Third Millennium,* ed. R. Subasinghe, P. Bueno, M. J. Phillips, C. Hough, S. E. McGladdery, and J. R. Arthur. Bangkok: NACA; Rome: FAO. Technical Proceedings of the Conference on Aquaculture in the Third Millennium, Bangkok, Thailand. February 20–25, 2000.

Delgado, C. L., N. Wada, M. W. Rosegrant, S. Meijer, and M. Ahmed. 2003. "Fish to 2020: Supply and Demand in Changing Global Markets." International Food Policy Research Institute, Washington, DC, and WorldFish Center, Penang, Malaysia.

Dey, M. M. 2000. "The Impact of Genetically Improved Farmed Tilapia in Asia." *Aquaculture Economics and Management* 4 (1-2): 107–24.

Dey, M. M., F. J. Paraguas, N. Srichantuk, Y. Xinhua, R. . Bhtta, and L. T. C. Dung. 2005. "Technical Efficiency of Freshwater Pond Polyculture Production in Selected Asian Countries: Estimation and Implications." *Aquaculture Economics and Management* 9 (1-2): 39–63.

Dey, M. M., M. A. Rab, K. M. Jahan, A. Nisapa, A. A. Kumar, and M. Ahmed. 2004. "Food Safety Standards and Regulatory Measures: Implications for Selected Fish Exporting Asian Countries." *Aquaculture Economics and Management* 8 (3-4): 217–36.

Dunham, R. A., K. Majumdar, E. Hallerman, D. Bartley, G. Mair, G. Hulata, Z. Liu, N. Pongthana, J. Bakos, D. Penman, M. Gupta, P. Rothlisberg, and G. Hoerstgen-Schwark. 2001. "Review of the Status of Aquaculture Genetics." In *Aquaculture in the Third Millennium*, ed. R. Subasinghe, P. Bueno, M. J. Phillips, C. Hough, S. E. McGladdery, and J. R. Arthur, 137–66. Bangkok: NACA; Rome: FAO.

The Economist. 2006. "The Omega Point." The Economist Newspaper and The Economist Group. January 19.

Edwards, P. 2000. "Aquaculture, Poverty Impacts and Livelihoods." *Natural Resources Perspectives* 56. June 2000. Available online at http://www. odi.org.uk/nrp/56 .html.

———. 2005. "Small-Scale Pond Aquaculture in Bangladesh." *Aquaculture Asia* X (4): 5–7.

EIFAC/FAO (European Inland Fisheries Advisory Commission/Food and Agriculture Organization). 1998. Report of the Symposium on Water for Sustainable Inland Fisheries and Aquaculture held in connection with the European Inland Fisheries Commission, 20th session, Praia do Caroveiro, Portugal. June 23–July 1. FAO Fisheries Report 580 (Suppl.). FAO, Rome.

Eurofish. 2004. "Does Farmed Salmon Cause Cancer?" 1: 62–65.

FAO (Food and Agriculture Organization of the United Nations). 1976. Report of the FAO Technical Conference on Aquaculture. Kyoto, Japan, May 26–June 2. FAO, Rome.

———. 1990. "CWP Handbook of Fishery Statistical Standards—Section J: Aquaculture." FIGIS Ontology Sheets. FAO, Rome. Available at: http://www.fao.org/figis/servlet/static?dom=ontology&xml=sectionJ.xml.

———. 1995a. "Precautionary Approach to Capture Fisheries and Species Introductions." FAO Technical Guidelines for Responsible Fisheries, No. 2. FAO, Rome.

———. 1995b. "Code of Conduct for Responsible Fisheries." FAO, Rome.

———. 1996. "Fisheries Department." FAO Technical Guidelines for Responsible Fisheries, No. 2. FAO, Rome.

———. 1997a. "Aquaculture Development." FAO Technical Guidelines for Responsible Fisheries, No. 5. FAO, Rome.

———. 1997b. "Review of the State of World Aquaculture." *FAO Fisheries Circular* No. 886, Rev 1. FAO, Rome.

———. 1999. "The State of World Fisheries and Aquaculture 1998." FAO, Rome.

———. 2000. "The State of World Fisheries and Aquaculture 2000." FAO, Rome.

———. 2003. "Asia-Pacific Aquaculture Sector Reviews: A Regional Synthesis." Unpublished Report. FAO, Rome.

———. 2004. "The State of Fisheries and Aquaculture 2004." FAO, Rome.

————. 2005a. "Regional Review on Aquaculture Development, 3. Asia and the Pacific." FAO Fisheries Circular 1017/3. FAO, Rome.

————. 2005b. "Regional Review of Aquaculture Development in Latin America and the Caribbean." Unpublished draft. FAO, Rome.

————. 2006. "Asia Pacific Regional Aquaculture Review—A Regional Synthesis." FAO, Rome.

————. In press. "Regional Review on Aquaculture Development. Africa." FAO Fisheries Circular. FAO, Rome.

————. In press. Draft Synthesis Report on Global Aquaculture. FAO Fisheries Department, Inland Water Resources, and Aquaculture Service (FIRI), Rome.

FAO Aquaculture Glossary. Available at http://www.fao.org/fi/glossary/aquaculture/ (accessed 2006).

FAO Fishstat. 2005. Available at http://www.fao.org/fi/statist/FISOFT/FISHPLUS.asp (databases downloaded May 2006).

FAO/NACA. 2000. "Asia Regional Technical Guidelines on Health Management for the Responsible Movement of Live Aquatic Animals and the Beijing Consensus and Implementation Strategy." FAO Fisheries Technical Paper 402. FAO, Rome.

————. 2001. "Manual of Procedures for the Implementation of the Asia Regional Technical Guidelines on Health Management for the Responsible Movement of Live Aquatic Animals." FAO Fisheries Technical Paper 402 (Suppl. 1). FAO, Rome.

FAO/NACA/CIFA. 2003. "Report of the FAO/NACA/CIFA Expert Consultation on the Intensification of Food Production in Low Income Food Deficit Countries through Aquaculture." Bhubaniswar, India, October 16–19, 2001. FAO Fisheries Report 718. FAO, Rome.

FAO/NACA/SEAFDEC/MRC/WFC. 2002. "The Role of Aquaculture and Living Aquatic Resources: Priorities for Support and Networking." Report of a regional donor consultation. Manila, November 27–28, 2002. FAO RAP, Bangkok.

FAO Regional Office for Africa. 1999. "Africa Regional Aquaculture Review." Proceedings of a Workshop held in Accra, Ghana, September 22–24, 1999. CIFA Occasional Paper 24. Accra: FAO.

————. 2005. Report of the Twenty-Sixth Session of the Committee on Fisheries. Rome, March 7–11, 2005. FAO Fisheries Report 780. FIPL/R780 (En). FAO, Rome.

FAO, UNDP, and Norway. 1987. "Thematic Evaluation of Aquaculture." Joint study by United Nations Development Programme, Norwegian Ministry of Development Cooperation and the Food and Agriculture Organization of the United Nations. Rome: FAO.

Feare, C. J. 2006. "Fish Farming and the Risk of Spread of Avian Influenza." Wild Wings Bird Management, March. Available at: http://www.birdlife.org/action/science/species/avian_flu/index.html.

Fegan, D. 1996. "Sustainable Shrimp Farming in Asia: Vision or Pipe Dream?" *Aquaculture Asia* 1 (2): 22–28.

Fiskeridirektoratet. 2004. "Lønnsomhetsundersøkelse for matfiskproduksjon Laks og Ørret. Omfatter selskaper med produksjon av laks og ørret i saltvann." Fiskeridirektoratet, Bergen.

FMC Biopolymer (USA) cited by I. Bryceson. 2005. Fishery Forum for Development Cooperation Annual Meeting, Bergen, 2005. (PowerPoint.)

Folke, C., N. Kautsky, H. Berg, Å. Jansson, and M. Troell. 1998. "The Ecological Footprint Concept for Sustainable Seafood Production—A Review." *Ecological Applications* 8 (1): 63–71.

Forster, J. 1999. "Aquaculture Chickens, Salmon: A Case Study." *World Aquaculture Magazine* 30 (3): 33, 35–38, 40, 69–70.

Freshwater Fisheries Research Centre (Wuxi, China). 2001. "Studies on the Recent Status of the Genetic Quality of the Major Chinese Farmed Fishes." Unpublished report. World Bank, Washington, DC.

Furedy, C. 1990. "Social Aspects of Human Excreta Reuse: Implications for Aquacultural Projects in Asia." In *Wastewater-Fed Aquaculture,* ed. P. Edwards and R. S. V. Pullin, 251–66. Proceedings of the International Seminar on Wastewater Reclamation and Reuse for Aquaculture, Calcutta, India, December 6–9, 1988. xxix. Environmental Sanitation Information Center, Asian Institute of Technology, Bangkok, Thailand.

Gjedrem, T. 1997. "Selective Breeding to Improve Aquaculture Production." *World Aquaculture* (March): 33–45.

———. 2000. "Genetic Improvement of Coldwater Fish." *Aquaculture Research* 31 (1): 25–33.

———. 2006. Personal communication.

Goldburg, R. J., M. S. Elliot, and R. L. Naylor. 2001. "Marine Aquaculture in the United States: Environmental Impacts and Policy Options." Arlington, VA: Pew Ocean Commission.

Gregussen, O. 2005. "Aquaculture—the Norwegian Experience." Presentation at the conference, "The Role of Aquaculture in Meeting Global Food Demand," Seattle, WA. June 20, 2005.

Griffiths, D. n.d. Poverty and Livelihoods: Field experiences from Bangladesh, Cambodia and Viet Nam. Available at http://www.streaminitiative.org/Library/pdf/pdf-FAO-Poverty/Don_Griffiths.pdf.

Gu, H. X., S. L. Hu, and Z. G. Yang. 1996. "The Relationship between Integrated Fish Farming and Human Influenza Pandemic." Zhonghua Liu Xing Bing Xue Za Zhi 17 (1): 29–32 (in Chinese).

Gupta, M. V., J. D. Sollows, M. A. Mazid, A. Rahman, M. G. Hussain, and M. M. Dey, ed. 1998. "Integrating Aquaculture with Rice Farming in Bangladesh: Feasibility and Economic Viability, Its Adoption and Impact." ICLARM Technical Report 55. ICLARM, Manila.

Hagler, M. 1997. "The Social Damage Caused by Shrimp Farming." In Shrimp: The Devastating Delicacy, A Greenpeace Report. Available at: http://archive.greenpeace .org/oceans/.

Hallman K., D. J. Lewis, and S. Begum. 2003. "An Integrated Economic and Social Analysis to Assess the Impact of Vegetable and Fishpond Technologies on Poverty in Rural Bangladesh." Food Consumption and Nutrition Division Discussion Paper 163. International Food Policy Research Institute, Washington, DC.

Halwart, M., and A. A. van Dam, ed. 2006. "Integrated Irrigation and Aquaculture in West Africa: Concepts, Practices and Potential." Rome, FAO.

Halwart, M., and M. V. Gupta, ed. 2004. Culture of Fish in Rice Fields. FAO and The WorldFish Center. Rome and Penang.

Hardy, R. W., and D. Gatlin. 2002. "Nutritional Strategies to Reduce Nutrient Losses in Intensive Aquaculture." In *Avances en Nutrición Acuícola VI,* ed. L. E. Cruz-Suárez, D. Ricque-Marie, M. Tapia-Salazar, M. Gaxiola-Cortés, and N. Simoes. Memorias del VI Simposium Internacional de Nutrición Acuícola. Septiembre 3–6, 2002. Cancún, Quintana Roo, México.

Harrison, E., J. A. Stewart, R. L. Stirrat, and J. Muir. 1994. *Fish Farming in Africa. What's the Catch?* London: Overseas Development Administration.

Hecht, T. 2006. "A Synopsis of Sub-Saharan African Aquaculture." A report prepared for the Fisheries Department Group (RAFI-FIRI) of the FAO Regional Office for Africa (draft). Grahamstown, South Africa.

Hishamunda, N., and R. Subasinghe. 2003. "Aquaculture Development in China: The Role of Public Policies. FAO Fisheries Technical Paper 427. FAO, Rome.

Hites, R. A., J. A. Foran, D. O. Carpenter, M. Coreen Hamilton, B. A. Knuth, S. J. Schwager. 2004. "Global Assessment of Organic Contaminants in Farmed Salmon." Science 303 (5655): 226–29.

Hop, L. T. 2003. "Program to Improve Production and Consumption of Animal Source Foods and Malnutrition in Vietnam." *Journal of Nutrition* 133: 4006S–4009S.

Hulata, G. 2001. "Genetic Manipulations in Aquaculture: A Review of Stock Improvement by Classical and Modern Technologies." *Genetica* 111 (1-3): 155–73.

ICES. 2003. Report of the Working Group on Marine Shellfish Culture, Trondheim, Norway. August 13–15. Mariculture Committee ICES CM 2003/F:05 Ref. ACME. ICES, Copenhagen.

———. 2005. Report of the Working Group on Environmental Interactions of Mariculture. ICES Mariculture Committee. April 11–15, 2005. Ottawa, Canada. CM 2005/F:04. Ref. I, ACME.

International Council for the Exploration of the Seas (ICES). 1995. "ICES Code of Practice on the Introduction and Transfers of Marine Organisms—1994." ICES Cooperative Research Report 204. ICES, Copenhagen.

International Finance Corporation Web site. 2006. Available at http://www.ifc .org/ifcext/agribusiness.nsf/Content/Aquaculture (accessed 2006).

International Foundation for Science Web site. 2006. Available at http://www.ifs.se/ (accessed 2006).

International Institute of Rural Reconstruction and International Center for Living Aquatic Resource Management, ed. 2001. "Integrated Agriculture-Aquaculture: A Primer." FAO Fisheries Technical Paper 407. FAO, Rome.

International Network on Genetics in Aquaculture Web site. 2006. Available at http:// www.worldfishcenter.org/inga/ (accessed 2006).

International Organization for Standardization. ISO 9000. Available at http://www.iso. org/iso/en/iso9000-14000/certification/index_two.html (accessed 2006).

Irz, X., J. R. Stevenson, A. Tanoy, P. Villarante, and P. Morissens. n.d. "Aquaculture and Poverty—A Case Study of Five Coastal Communities in the Philippines." Research project R8288 DFID AFGRP: Assessing the Sustainability of Brackish Water Aquaculture Systems in the Philippines. Working Paper 4.

Islam, M. S., M. A. Wahab, A. A. Miah, and A. H. M. Mustafa Kamal. 2003. "Impacts of Shrimp Farming on the Socioeconomic and Environmental Conditions in the Coastal Regions of Bangladesh." *Pakistan Journal of Biological Sciences* 6 (24): 2058–2067.

Josupeit, H. 1985. "A Survey of External Assistance to the Fisheries Sector in Developing Countries, 1978–1984." FAO Fisheries Circular 755. FAO, Rome.

Josupeit, H., A. Lem, and H. Lupin. 2001. "Aquaculture Products: Quality, Safety, Marketing and Trade." In *Aquaculture in the Third Millennium,* ed. R. Subasinghe, P. Bueno, M. J. Phillips, C. Hough, S. E. McGladdery, and J. R. Arthur. Bangkok: NACA; Rome: FAO. Technical Proceedings of the Conference on Aquaculture in the Third Millennium, Bangkok, Thailand. February 20–25, 2000.

Juniati. 2005. "An Integrated Livestock-Fish Farming System." *LEISA Magazine on Low External Input and Sustainable Agriculture* 21 (3): 29.

Kapetsky, J. M. 1994. "A Strategic Assessment of Warm-Water Fish Farming Potential in Africa." CIFA Technical Paper 27. FAO, Rome.

———. 1995. A First Look at the Potential Ccontribution of Warm Water Fish Farming to Food Security in Africa. In *Proceedings of the Seminar on the Management of Integrated Freshwater Agro-Piscicultural Ecosystems in Tropical Areas,* ed. J.-J. Symoens and J.-C. Micha, 547–592. Brussels, Belgium, May 16–19, 1994. Technical Centre for Rural Cooperation (CTA), Wageningen, the Netherlands.

Kapuscinski, A. R. 2005. "Current Scientific Understanding of the Environmental Biosafety of Transgenic Fish and Shellfish." *Rev. Sci. SciTech. Off. Int. Epiz.* 2005, 24 (1): 309–322.

Karim, M., M. Ahmed, R. K. Talukder, M. A. Taslim, and H. Z. Rahman. 2006. "Dynamic Agribusiness-Focused Aquaculture for Poverty Reduction and Economic Growth in Bangladesh." WorldFish Center Discussion Series Policy Working Paper. 1.44. Ministry of Fisheries and Livestock, Bangladesh, The WorldFish Center, and Bangladesh Shrimp and Fish Foundation, Penang.

Kaspar, H. F., P. A. Gillespie, I. C. Boyer, and A. L. MacKenzie. 1985. "Effects of Mussel Aquaculture on the Nitrogen Cycle and Benthic Communities in Kenepuru Sound, Marlborough Sounds, New Zealand." *Marine Biology* 85: 127–36.

Kelleher, K. 1974. "An Analysis and Evaluation of Subsistence-Level Tilapia Culture in the Central African Republic." Project report. FAO UNDP Bangui.

———. 2005. "Discards in the World's Marine Fisheries. An Update." FAO Fisheries Technical Paper 470. FAO, Rome.

King, H. R. 1993. "Aquaculture Development and Environmental Issues in Africa." In *Environment and Aquaculture in Developing Countries,* ed. R. S. V. Pullin, H. Rosenthal, and J. L. Maclean, 116–124. ICLARM Conference Proceedings 31, WorldFish Center, Penang, Malaysia.

King, J. 1992. "Aquaculture in southern Africa." Food and Agriculture Organization of the United Nations, Harare, Zimbabwe.

Konde, V. 2006. "Africa in the Global Flows of Technology: An Overview." African Technology Development Forum (ATDF) Journal 3 (1). Available at www.atdforum.org (accessed 2006).

Kongkeo, H. 2001. "Current Status and Development Trends of Aquaculture in the Asian Region." In *Aquaculture in the Third Millennium,* ed. R. Subasinghe, P. Bueno, M. J. Phillips, C. Hough, S. E. McGladdery, and J. R. Arthur, 267–93. Bangkok: NACA; and Rome: FAO. Technical Proceedings of the Conference on Aquaculture in the Third Millennium, Bangkok, Thailand, February 20–25, 2000.

Kumaran, M., N. Kalaimani, K. Ponnusamy, V. S. Chandrasekaran, and D. Deboral Vimala. 2003. "A Case of Informal Shrimp Farmers' Association and Its Role in Sustainable Shrimp Farming in Tamil Nadu." *Aquaculture Asia* 7 (2): 10–12.

Lam, T. D. T. 2003. "US 'catfish war' defeat stings Vietnam." *Asia Times* online. July 31.

Le, Than Luu. 2001. "Sustainable Aquaculture for Poverty Alleviation (SAPA): A New Rural Development Strategy for Viet Nam—Part I." *FAO Aquaculture Newsletter* 27. July.

Lewis, D. J., G. D. Wood, and R. Gregory. 1996. *Trading the Silver Seed, Local Knowledge and Market Moralities in Aquaculture Development.* Dhaka: University Press Limited.

Li, S. F. 2003. "Aquaculture Research and Its Relationship to Development in China." In Agricultural Development and the Opportunities for Aquatic Resources Research in China. WorldFish Center, Penang.

Little, D. C., and P. Edwards. 2003. "Integrated Livestock-Fish Farming Systems." Inland Water Resources and Aquaculture Service. Animal Production Service. FAO, Rome.

LMC International Ltd. n.d. As cited in A. G. J. Tacon, "State of Information on Salmon Aquaculture Feed and the Environment." Available at http://www.westcoast aquatic.ca/Aquaculture_feed_environment.pdf (accessed 2).

Lowther, A. 2005. "Highlights from the FAO Database on Aquaculture Production Statistics." *FAO Aquaculture Newsletter* 33 (July): 22–24. FAO, Rome.

Luu, L. T. 1992. "The VAC System in Northern Vietnam." In: FAO/ICLARM/IIRR. Integrated Agriculture-Aquaculture: A Primer. FAO Fisheries Technical Paper 407. Rome, FAO. 2001.

MacAlister, Elliott, and Partners. 1999. "Forward Study of Community Aquaculture. Summary Report." European Commission, Fisheries Directorate General.

Mair, G. C., and P. A. Tuan. 2002. "Vietnam: Stock Comparisons for Polyculture and National Breeding Programmes." In proceedings of a workshop on "Genetic Management and Improvement Strategies for Exotic Carps in Asia." ed. D. J. Penman, M. G. Hussain, B. J. McAndrew, and M. A. Mazid, 37–42. Dhaka, Bangladesh. February 12–14, 2002. Bangladesh Fisheries Research Institute, Mymensingh, Bangladesh.

Mandal, S. A., G. Chowhan, G. Sarwar, A. Begum, and A. N. M. Rokon Uddin. 2004. Mid-term Review Report on Development of Sustainable Aquaculture Project CDSAP. Dhaka, Bangladesh: WorldFish Center, South Asia Office.

Mazid, M., ed. 1999. Rural and Coastal Aquaculture in Poverty Reduction. Proceedings of a seminar organized on the occasion of Fish Week 1999. Ministry of Fisheries and Livestock, Bangladesh.

Moehl, J. 2003. "Africa Regional Activities." *FAO Aquaculture Newsletter* 29 (July).

Moehl, J., M. Halwart, and R. Brummett. 2005. "Report of the FAO-WorldFish Center Workshop on Small-Scale Aquaculture in Sub-Saharan Africa: Revisiting the Aquaculture Target Group Paradigm." Limbé, Cameroon, March 23–26, 2004. CIFA Occasional Paper 25. FAO, Rome.

Monfort, M. C. 2006. "Markets and Marketing of Aquaculture Finfish in Europe Focus on the Mediterranean Basin." FAO Fisheries Division. April 2006.

Monti, G., and M. Crumlish. n.d. "The Importance of Pangasius Farming in the Mekong Delta, Vietnam." AquaNews. Available at http://www.aquaculture.stir.ac/uk/AquaNews/32P18_20.pdf (accessed 2006).

Myrseth, B. In press. *What We Have Learned from Fish Farming and How Can We Apply This for Future Developments.* Norway: Marine Farms ASA.

NACA/FAO, ed. 2001. "Aquaculture Development: Financing and Institutional Support." In *Aquaculture in the Third Millennium,* ed. R. Subasinghe, P. Bueno, M. J. Phillips, C. Hough, S. E. McGladdery, and J. R. Arthur. Bangkok: NACA; Rome: FAO.

Nairobi Declaration on Conservation of Aquatic Biodiversity and Use of Genetically Improved and Alien Species for Aquaculture in Africa. 2002. ICLARM/CTA/ FAO/ IUCN/UNEP/CBD. Nairobi, Kenya.

Nash, C. E., P. R. Burbridge, and J. K. Volkman, ed., 2005. "Guidelines for Ecological Risk Assessment of Marine Aquaculture." Prepared at the NOAA Fisheries Service Manchester Research Station International Workshop, April 11–14, 2005. U.S. Department of Commerce, National Oceanic and Atmospheric Agency (NOAA), National Marine Fisheries Service. Tech. Memo. NMFS-NWFSC-71.

National Research Council (NRC). 1999. "Clean Coastal Waters: Understanding and Reducing the Effects of Nutrient Pollution." National Academy Press, Washington, DC.

Naylor, R., R. J. Goldburg, H. Mooney, M. Beveridge, J. Clay, C. Folke, N. Kautsky, J. Lubchenco, J. Primavera, and M. Williams. 1998. "Nature's Subsidies to Shrimp and Salmon Farming." *Science* 282 (5390): 883.

Naylor, R., K. Hindar, I. A. Fleming, R. Goldburg, S. Williams, J. Volpe, F. Whoriskey, J. Eagle, D. Kelso, and M. Mangel. 2005. "Fugitive Salmon: Assessing the Risks of Escaped Fish from Net-Pen Aquaculture." *BioScience* 55 (5): 427–37.

Naylor, R. L., R. J. Goldburg, J. H. Primavera, N. Kautsky, M. C. M. Beveridge, J. Clay, C. Folke, J. Lubchenco, H. Mooney, and M. Troell. 2000. "Effects of Aquaculture on World Fish Supplies." *Nature* 405: 1017–024.

Network of Aquaculture Centers in Asia-Pacific (NACA). 2006. "Thailand, Shrimp Farming and the Environment." Available at http://www.enaca.org/modules/tinyd2/index.php?id=1 (accessed 2006).

OIE. 2005. Aquatic Animal Health Code, 8th edition. Available at http://www.oie.int/eng/normes/fcode/A_summry.htm (accessed April 2006).

Olesen, I., T. Gjedrem, H. B. Bentsen, B. B. Gjerde, and M. Rye. 2003. "Breeding Programs for Sustainable Aquaculture." *Journal of Applied Aquaculture* 13 (3-4): 179–204.

Oswald, M., Y. Copin, and D. Montferrer. 1996. "Peri-urban Aquaculture in Midwestern Côteô d'Ivoire." *ICLARM Conf. Proc.* 41: 525–536.

Ottolenghi, F., C. Silvestri, P. Giordano, A. Lovatelli, and M. B. New. 2004. "Capture-Based Aquaculture, the Fattening of Eels, Groupers, Tunas and Yellowtails." FAO, Rome.

Pauly, D., E. Froese, L. Y. Liu, and P. Tyedmers. 2001. "Down with Fisheries, Up with Aquaculture: Implications of Global Trends in the Mean Trophic Levels of Fish." Paper presented at American Association for the Advancement of Science-sponsored mini-symposium, The Aquaculture Paradox: Does Fish Farming Supplement or Deplete World Fisheries, San Francisco, California. February 18, 2001.

Penamn, D. J. 2005. "Progress in Carp Genetics Research." In *Carp Genetic Resources for Aquaculture in Asia*, ed. D. J. Penman, M. V. Gupta, and M. M. Dey. WorldFish Center Technical Report 65. WorldFish Center, Penang.

Phillips, M. J., M. C. M. Beveridgand, and R. M. Clarke. 1991. "Impact of Aquaculture on Water Resources." In *Advances in Aquaculture* 3, ed. D. E. Brune and J. R. Tomasso, 568–91. Baton Rouge, LA: World Aquaculture Society.

Pillay, T. V. R. 2001. "Aquaculture Development: From Kyoto 1976 to Bangkok 2000." In *Aquaculture in the Third Millennium,* ed. R. Subasinghe, P. B. Bueno, M. J. Philipps, C. Hough, S. E. McGladdery, and J. R. Arthur. Bangkok: NACA; Rome: FAO.

Plan of Implementation of the World Summit on Sustainable Development. 2002. Available at: http://www.un.org/esa/sustdev/documents/WSSD_POI_PD/English/POIToc.htm.

Pongthanapanich, T. 2005. "Thai Shrimp Farming. How Much Should Be Taxed?" PowerPoint presentation at the FAME workshop, University of Southern Denmark.

Radheyshyam, M. 2001. "Community-Based Aquaculture in India—Strengths, Weaknesses, Opportunities, Threats." *Naga: The ICLARM Quarterly* 24 (1-2).

Rahim, M. R. 2004. "Profitability of Some Newly Introduced Rice Varieties in Selected Locations of Bangladesh." M.S. thesis. Department of Agricultural Economics, Bangladesh Agricultural University, Mymensingh, Bangladesh.

Rahman, H. Z., M. Nuruzzaman, S. Z. Ahmed, M. Z. Rahman, and M. B. Hossain. 2005. "The Interface of Community Approaches and Agri-Business: Insights from Floodplain Aquaculture in Daudkandi." Draft final report. Power and Participation Research Centre, House 79, Road 12/A, Dhanmondi, Dhaka 1209, Bangladesh.

Rana, K., J. Anyila, K. Salie, C. Mahika, S. Heck, and J. Young. n.d. "The Role of Aqua Farming in Feeding African Cities." *Urban Agriculture* 14: 36–38.

Rana, K., J. Anyila, K. Salie, C. Mahika, S. Heck, J. Young, and G. Monti. 2005. "Aquafarming in Urban and Peri-urban Zones in Sub-Saharan Africa." Presentation at the 7th Bi-annual Conference of the Aquaculture Association of southern Africa, Grahamstown. September 12–14.

Redding, T. A., and A. B. Midlen. 1990. "Fish Production in Irrigation Canals. A Review." FAO Fisheries Technical Paper 317. FAO, Rome.

Rijsberman, F. R., ed. 2000. "World Water Scenarios: Analyses." Draft. Earthscan, London. Feb. 25.

Roderick, E. 2002. "Food of Kings Now Feeding the Masses!" *Fish Farming International,* 29 (9): 32–34.

Satia, B. P., P. N. Satia, and A. Amin. 1992. "Large-scale Reconnaissance Survey of Socioeconomic Conditions of Fish Farmers and Aquaculture Practices in the West and Northwest Provinces of Cameroon." Aquaculture Research Systems in Africa. 64–90. IDRC-MR308e.f., International Development Research Centre, Ottawa, Ontario, Canada.

Scholz, U., and S. Chimatiro. 1996. "The Promotion of Small-Scale Aquaculture in the Southern Region of Malawi, a Reflection of Extension Approaches and Technology Packages Used by the Malawi-German Fisheries and Aquaculture Development Project (MAGFAD)." In Vieh und Fisch. Fachliche Beiträge über Viehwirtschaft, Veterinärwesen und Fischerei aus Projekten und der Zentrale. GTZ Publication: Eschborn, Germany.

Shehadeh, Z. H., and J. Orzeszk. 1997. "External Assistance." In Review of the State of World Aquaculture. FAO Fisheries Circular 886. FAO, Rome.

Shepherd, J. 2005. "Sustainability and World Market Prospects." IFFO presentation. Learning Conference on the Sustainability of Peruvian Industrial Anchoveta Industry, Lima, Peru. August 31–September 2.

Shilu, L., and H. Linhau. 1999. "Key Issues in Aquaculture Development in China in the 21st Century." INFOFISH International 1: 42–45.

Simpson, B. 2006. "The Transfer and Dissemination of Agricultural Technologies: Issues, Lessons and Opportunities." African Technology Development Forum (ATDF) Journal 3 (1): 10–17. Available at http://www.atdforum.org (accessed 2006).

Skladany, M. 1996. "Fish, Pigs, Poultry and Pandora's Box: Integrated Aquaculture and Human Influenza." In *Aquaculture Development, Social Dimensions of an Emerging Industry,* ed. C. Bailey, S. Jentoft, and P. Sinclair, 267–85. Boulder, CO: Westview Press.

Smoker, W. W. 2004. "Regional Non-Profit Corporations—An Institutional Model for Stock Enhancement." In *Stock Enhancement and Sea Ranching. Developments, Pitfalls and Opportunities,* ed. K. M. Leber, S. Kitada, T. Svasand, and H. L. Blankenship. Oxford, U.K.: Blackwell.

STREAM [Support to Regional Aquatic Resources Management] Initiative. 2005. "One-Stop Aqua Shops" Better-Practice Guidelines No. 19.

Subasinghe, R. P., and M. J. Phillips. 2002. "Aquatic Animal Health Management: Opportunities and Challenges for Rural, Small-Scale Aquaculture and Enhanced Fisheries Development: Workshop Introductory Remarks." In *Primary Aquatic Animal Health Care in Rural, Small-Scale Aquaculture Development,* ed. J. R. Arthur, M. J. Phillips, R. P. Subasinghe, M. B. Reantaso, and I. H. MacRae, 1–5. FAO Fisheries Technical Paper 406. FAO, Rome.

Subasinghe, R. P., J. R. Arthur, and M. Shariff, ed. 1996. "Health Management in Asian Aquaculture." Proceedings of the Regional Expert Consultation on Aquaculture Health Management in Asia and the Pacific Serdang, Malaysia. May 22–24, 1995. FAO Fisheries Technical Paper 360. FAO, Rome.

Subasinghe, R. P., M. G. Bondad-Reantaso, and S. E. McGladdery. 2001. "Aquaculture Development, Health and Wealth." In *Aquaculture in the Third Millennium*, ed. R. Subasinghe, P. Bueno, M. J. Philips, C. Hough, S. E. McGladdery, and J. R. Arthur, 167–91. Bangkok: NACA; Rome: FAO. Technical Proceedings of the Conference on Aquaculture in the Third Millennium, Bangkok, Thailand. February 20–25, 2000.

Support to Regional Aquatic Resources Management (STREAM) Initiative. 2005. Available at http://www.streaminitiative.org/Library/pdf/bpg/worlp/19WBPG.pdf (accessed 2006).

Sustainable Aquaculture for Poverty Alleviation (SAPA). 2000. Proceedings of the Scoping Meeting for Development of the Sustainable Aquaculture for Poverty Alleviation (SAPA), Ministry of Fisheries, Hanoi. May 23–25.

Swick, R. A., and M. C. Cremer. 2001. "Livestock Production: A Model for Aquaculture?" In *Aquaculture in the Third Millennium*, ed. R. Subasinghe, P. Bueno, M. J. Philips, C. Hough, S. E. McGladdery, and J. R. Arthur, 49–60. Bangkok: NACA; Rome: FAO.

Szuster, B. W. 2003. "Shrimp Farming in Thailand's Chao Prayha Delta." Pelawatte, Sri Lanka: International Water Management Institute.

Tacon, A. G. J. 2003. "Global Trends in Aquaculture and Compound Aquafeed Production: A Review." International Aquafeed Directory and Buyers' Guide 8–23.

———. 2005. "Salmon Aquaculture Dialogue. State of Information on: Salmon Aquaculture Feed and the Environment." La Sostenibilidad de la Pesca Industrial de la Anchoveta en el Perú. Lima, Peru. PowerPoint presentation.

———. 2006. "Trash Fish Fisheries, Aquaculture, Pellets and Fishmeal Substitutes." Asia-Pacific Fisheries Commission. Regional Consultative Forum, Meeting, Kuala Lumpur. August.

Talukder, R. K. 2004. "Socioeconomic Profiles of the Stakeholders of the Aquaculture Sector in Bangladesh." Paper presented at the Final Workshop on Strategies and Options for Increasing and Sustaining Fisheries and Aquaculture Production to Benefit Poor Households in Bangladesh, Manila, Philippines. March 17–20.

Thompson, P., P. Sultana, and A. K. M. Firoz Khan. 2005. "Aquaculture Extension Impacts in Bangladesh: A Case Study from Kapasia, Gazipur." WorldFish Center Technical Report 63. WorldFish Center, Penang.

U.S. Agency for International Development (USAID) SPARE Fisheries and Aquaculture Panel. n.d. "Review of the Status, Trends and Issues in Global Fisheries and Aquaculture, with Recommendations for USAID Investments." Available at http://pdacrsp.oregonstate.edu/miscellaneous/F%26A_Subsector_Final_Rpt.pdf (accessed 2006).

U.S. Department of Commerce. 2005. "Guidelines for Ecological Risk Management of Marine Fish Aquaculture." NOAA Technical Memorandum NMFS-NWFSC-71. NOAA, Silver Spring, MD.

USINFO (International Information Programs) Web site. 2004. Available at http://usinfo.state.gov/ (accessed 2006).

VACVINA. 1995. "Intensive Small-Scale Farming in Vietnam." ILEIA Newsletter 11 (1): 4. Available at http://www.metafro.be/leisa/1995/11-1-4.pdf (accessed 2006).

Van der Mheen, H. 1998. "Achievements of Smallholder Aquaculture Development in Southern Africa." Experiences from the ALCOM Aquaculture Programme. FAO, Harare, Zimbabwe.

Walker, P. J. 2004. "Disease Emergence and Food Security: Global Impact of Pathogens on Sustainable Aquaculture Production." In *Fish, Aquaculture and Food Supply: Sustaining Fish as a Food Supply,* ed. I. Brown, 44–50. Record of a conference conducted by the ATSE Crawford Fund, Canberra, Australia. August 11.

Wang, Y. 2001. "China P. R.: A Review of National Aquaculture Development." In *Aquaculture in the Third Millennium,* ed. R. Subasinghe, P. Bueno, M. J. Phillips, C. Hough, S. E. McGladdery, and J. R. Arthur, 307–16. Bangkok: NACA; Rome: FAO. Technical Proceedings of the Conference on Aquaculture in the Third Millennium, Bangkok, Thailand. February 20–25, 2000.

Weber, J. T., E. D. Mintz, R. Canizares, A. Semiglia, I. Gomez, R. Sempertegui, A. Davila, K. D. Greene, N. D. Puhr, D. N. Cameron, F. C. Tenover, T. J. Barrett, N. H. Bean, C. Ivey, R.V. Tauxe, and P. A. Blake. 1994. "Epidemic Cholera in Ecuador: Multidrug-Resistance and Transmission by Water and Seafood." *Epidemiology and Infection* (112): 1–11.

Weinstein M. R., M. Litt, D. A. Kertesz, P. Wyper, D. Ross, M. Coulter, A. McGreer, · R. Facklam, C. Ostach, B. M. Willey, A. Borczyk, D. E. Low, and the Investigative Team. 1997. "Invasive Infection Due to a Fish Pathogen: Streptococcus iniae." *New England Journal of Medicine* 33 (7): 589–94.

WHO [World Health Organization]. 1999. "Food Safety Issues Associated with Product from Aquaculture, Report of a joint FOA/NACA/WHO study group. Geneva.

Wijkstrom, U. N. 2003. "Some Long-Term Prospects for Consumption of Fish." *Veterinary Research Communications 27* (Suppl. 1): 461–68.

Wilder, M., and N. T. Phuong. 2002. "The Status of Aquaculture in the Mekong Delta Region of Vietnam: Sustainable Production and Combined Farming Systems." Proceedings of International Commemorative Symposium: 70th Anniversary of the Japanese Society of Fisheries Science. *Fisheries Science 68* (Suppl. I).

World Bank, Center for Development and Integration (CDI), and Vietnam Institute of Economics (VIE). 2006. "Vietnam: Engagement of Poor Fishing Communities in the Identification of Resource Management and Investment Need." The World Bank and CDI in collaboration with VIE. June.

World Bank. 2005. "Technology and Growth Series: Chilean Salmon Exports." PREM Notes 103. Available at http://www.enaca.org/shrimp (accessed 2006).

———. 2006. "Meeting the Challenge of Africa's Development: A World Bank Group Action Plan." Available at http://siteresources.worldbank.org/INTAFRICA/Resources/aap_final.pdf (accessed 2006).

———. 2006. "Agriculture Investment Sourcebook" Available at: http://siteresources.worldbank.org/EXTAGISOU/Resources/Module6_Web.pdf. (accessed 2006).

World Organization for Animal Health (OIE). 2003. Aquatic Animal Health Code, sixth ed. Office International des Epizooties, Paris. Available at http://www.oie.int/eng/normes/ fcode/a_summary.htm (accessed 2006).

Xiuzhen, F. 2003. "Rice-Fish Culture in China." *Aquaculture Asia* 8 (4): 44–46.

Yap, W. G. 2004. "Policies and Strategies in the Commercialization of Aquaculture Development in SE Asia." FAO/NACA/SEAFDEC Aquaculture Department, Ilo-Ilo.

Ye, Y. 1999. "Historical Consumption and Future Demand for Fish and Fishery Products: Exploratory Calculations for the Years 2015–2030." FAO Fisheries Circular 946, FAO, Rome.

Zweig, R. 2006. Personal communication (e-mail).

SELECTED BIBLIOGRAPHY

Ahmed, M., M. A. Rab, and M. A. P. Bimbao. 1995. "Aquaculture Technology Adoption in Kapasia Thana Bangladesh: Some Preliminary Results from Farm Record-Keeping Data." ICLARM Technical Report 44. International Center for Living Aquatic Resources Management, Manila, Philippines.

Ali, A. B. 1992. "Rice-Fish Integration in Malaysia: Present Status and Future Prospects." In proceedings of the FAO/IPT Workshop on Integrated Livestock-Fish Productions Systems, "Integrated Livestock-Fish Production Systems," Kuala Lumpur, Malaysia. December 16–20, 1991.

Ayyapan, S. 1999. "Status and Role of Aquaculture in Rural Development in India." Paper presented at the FAO/NACA Consultation on Aquaculture for Sustainable Rural Development, Changrai, Thailand. March 29–31.

Boyd, C. E., and A. Gross. 2000. "Water Use and Conservation for Inland Aquaculture Ponds." *Fisheries Management and Ecology* 7 (1-1): 55–63.

Brugere, C., and N. Ridler. 2004. "Global Aquaculture Outlook in the Next Decades: An Analysis of National Aquaculture Production Forecasts to 2030." FAO Fisheries Circular 1001. FAO, Rome.

Brummett, R. E., and B. A. Costa-Pierce. 2002. "Village-Based Aquaculture Ecosystems as a Model for Sustainable Aquaculture Development in Sub-Saharan Aquaculture." In *Ecological Aquaculture: The Evolution of the Blue Revolution,* ed. B. A. Costa-Pierce, 145–60. Oxford, UK: Blackwell Science.

Bueno, P. 1999. "Small-Scale Aquaculture in Rural Development: Issues, Directions and Lessons." Unpublished. A synthesis of national reports presented at the FAO/NACA expert consultation on sustainable aquaculture for rural development, Chieng Rai, Thailand. March 29–31.

Funge-Smith, S., and M. J. Phillips. 2001. "Aquaculture Systems and Species." In *Aquaculture in the Third Millennium,* ed. R. Subasinghe, P. Bueno, M. J. Phillips, C. Hough, S. E. McGladdery, and J. R. Arthur, 129–35. Bangkok: NACA; Rome: FAO. Technical Proceedings of the Conference on Aquaculture in the Third Millennium, Bangkok, Thailand. February 20–25, 2000.

Morshed, S. M. 2004. Community Based Aquaculture in Flood Plains Implementation Guidelines (in Bangla), trans. Muhammad Muzaffar Hussain. Dhaka: SHISUK.

Muir, J. F. 1995. "Aquaculture Development Trends: Perspectives for Food Security." KC/FI/95/TECH/4. Paper presented at the International Conference on Sustainable Contribution of Fisheries to Food Security, Kyoto, Japan. December 4–9, 1995.

National Committee for Research Ethics in Science and Technology, Norwegian Academy of Sciences and Centre for Technology and Culture. 1998. Holmenkollen Guidelines for Sustainable Aquaculture (adopted 1998). Trondheim.

New, M., A. G. J. Tacon, and I. Csavas, ed. 1994. "Farm-made Aquafeeds." FAO Fisheries Technical Paper 343. FAO, Rome.

Phillips, M. J. 1998. "Tropical Mariculture and Coastal Environmental Management." In *Tropical Mariculture,* ed. S. S. De Silva, 38–69. New York: Academic Press.

Purwanto, E. 1999. "Status and Role of Aquaculture in Rural Development in Indonesia." Paper presented at the FAO/NACA Consultation on Aquaculture for Sustainable Rural Development, Changrai, Thailand. March 29–31.

Riggs, F. 1996. "Consumer Awareness of Environmental Protection Needs in the Shrimp Culture Industry: Marketing for Sustainability." Paper presented at World Aquaculture 1996, World Aquaculture Society, Bangkok, Thailand. January 29–February 2.

Rijsberman, F., N. Manning, and S. de Silva. 2006. "Increasing Green and Blue Water Productivity to Balance Water for Food and Environment." 4th World Water Forum, Mexico, March 2006. Food and Environment baseline paper. Available at http://www.iwmi.cgiar.org/WWF4/html/action_1.htm (accessed 2006).

Roberts, R. J., and J. F. Muir. 1995. "25 Years of World Aquaculture Sustainability, a Global Problem." In *Sustainable Fish Farming, Proceeding of the First International Symposium on Sustainable Fish Farming,* ed. H. Reinertsen and H. Haaland, 167–81. Rotterdam: A. A. Balkema. First International Symposium on Sustainable Fish Farming, Oslo, Norway. August 28–31.

Ruddle, K. 1993. "The Impacts of Aquaculture Development on Socioeconomic Environments in Developing Countries: Toward a Paradigm for Assessment." In *Environment and Aquaculture in Developing Countries,* ed. R. S. V. Pullin, H. Rosenthal, and J. L. Maclean, 20–41. ICLARM Conf. Proc. 31.

Stickney, R. R., and J. P. McVey. 2002. *Responsible Marine Aquaculture.* Wallingford: CABI Publishing.

Tacon, A. G. J. 2001. "Increasing the Contribution of Aquaculture for Food Security and Poverty Alleviation." In *Aquaculture in the Third Millennium,* ed. R. Subasinghe, P. Bueno, M. J. Philips, C. Hough, S. E. McGladdery, and J. R. Arthur, 63–72. Bangkok: NACA; Rome: FAO. Technical Proceedings of the Conference on Aquaculture in the Third Millennium, Bangkok, Thailand. February 20–25, 2000.

Tidwell, J. H., and G. L. Allan. 2001. "Fish as Food: Aquaculture's Contribution." *EMBO Reports* 21 (11): 103–08.

Turner, G., ed. 1988. "Codes of Practice and Manual of Procedures for Consideration of Introductions and Transfers of Marine and Freshwater Organisms." EIFAC Occasional Paper 23. European Inland Fisheries Advisory Commission. FAO, Rome.

Watanabe, T. 1993. The Ponds and the Poor. The Story of Grameen Bank's Initiative. Dhaka: Grameen Bank.

INDEX

The letters *b, f, t,* and n indicate boxes, figures, tables, and notes.

Meghna-Dhanagoda Command Area Development Project, 122*b*
Mekong delta, 60, 117
Métro, 36*t*
Mexico, 16*t*, 106*t*, 138*t*
Millennium Development Goals (MDGs), 56
models, 58, 77–8
mollusks, 155
monitoring, 90
Morocco, 144*t*
movement, 128*b*
mussel rearing, ecological footprint, 148*t*
Myanmar, 136*t*, 137*t*

Nairobi Declaration on Conservation of Aquatic Biodiversity, 72
Nantong Wangfu Special Quatic Products Co. Ltd., 108*t*
national codes, 99–100
national development plans, 75
national strategies, 5, 48–9
Netherlands, the, 138*t*, 147
Network of Aquaculture Centers in Asia-Pacific (NACA), 10, 51, 54, 129, 131, 133
Network of Aquaculture Centres in Central and Eastern Europe, 164n
networks, 77–8, 91, 129, 131, 134
New Zealand, 138*t*
Niger, 150
Nigeria, 69, 70, 138*t*, 144*t*, 151
Nile perch, 27
Noakhali Rural Development Program, 120
nongovernmental organizations (NGOs), 134
nontariff barriers (NTBs), 66, 67
Northwest Fisheries Extension Project, 120, 121*b*
Norway, 16*t*, 24, 32, 137*t*
    salmon, 9, 34*f*, 143*f*
Norwegian Salmon National Breeding Program, 43
Nova Companies Ltd., 108*t*
nurseries, 95
nutrients and wastes, 37

nutritional benefits of fish consumption, 31, 32*b*
nutritional value of fish, 42

OIE. *See* World Organization for Animal Health
omega-3, 32*b*
One-Stop Aqua Shops (OASs), 10, 52*b*, 55
oysters, 33*b*, 44*t*

Pacific white leg shrimp, 41*t*
Pan Fish, 14
paralytic shellfish poisoning, 31
Partnership for Development in Kampuchea, 134
partnerships, 11–2, 89, 92
pathogens, 30
pesticides, 31
Philippines, the, 64, 99, 122–3, 127*f*, 136*t*, 137*t*
    Fisheries Credit Project, 106*t*
Pilot Fishery Development Project, 104*t*
piscicides, 31
Poland, production rank, 138*t*
policies, 4–5, 65, 92, 113
    development, 75–6
    interventions, 123–6
    pro-poor, 60, 64
    protecting wild stock, 28
polyculture, 60, 95, 121*b*
pond fish culture, 118
pond-vegetable garden, 150
ponds, 95
poorest, the, 63, 120, 121–2, 121*b*, 122*b*
poverty impacts, 56–8
poverty reduction, 29, 60, 62, 64, 82
    and trade, 66–7
    Bangladesh, 118, 125
    China, 112
    pro-poor approaches, 77, 89
pregnancy, 32*b*
prices, 19, 40, 41*t*, 141*f*
private fingerling producers (PPAs), 150
private sector, 6, 8, 76, 87, 110, 160
pro-poor approaches, 7, 58, 60–2, 77, 89, 91, 118, 123–6